Anna Tramontano
Protein Structure Prediction

Related Titles

Eidhammer, I., Jonassen, I., Taylor, W. R.

Protein Bioinformatics
An Algorithmic Approach to Sequence and Structure Analysis

376 pages
2004
Hardcover
ISBN 0-470-84839-1

Höltje, H.-D., Sippl, W., Rognan, D., Folkers, G.

Molecular Modeling
Basic Principles and Applications

240 pages with 66 figures and 20 tables
2003
Softcover
ISBN 3-527-30589-0

Bourne, P. E., Weissig, H. (eds.)

Structural Bioinformatics

648 pages
2003
Softcover
ISBN 3-471-20199-5

Budis, N.

Engineering the Genetic Code
Expanding the Amino Acid Repertoire for the Design of Novel Proteins

312 pages with 76 figures and 7 tables
2006
Hardcover
ISBN 3-527-31243-9

Anna Tramontano
Protein Structure Prediction

Concepts and Applications

WILEY-VCH

WILEY-VCH Verlag GmbH & Co. KGaA

The Author

Prof. Dr. Anna Tramontano
University of Rome "La Sapienza"
Department of Biochemical Sciences "Rossi Fanelli"
P. le Aldo Moro, 5
00185 Rome
Italy

All books published by Wiley-VCH are carefully produced. Nevertheless, authors, editors, and publisher do not warrant the information contained in these books, including this book, to be free of errors. Readers are advised to keep in mind that statements, data, illustrations, procedural details or other items may inadvertently be inaccurate.

Library of Congress Card No.:
applied for

British Library Cataloguing-in-Publication Data
A catalogue record for this book is available from the British Library.

**Bibliographic information published by
Die Deutsche Bibliothek**
Die Deutsche Bibliothek lists this publication in the Deutsche Nationalbibliografie; detailed biliographic data is available in the Internet at <http://dnb.ddb.de>.

© 2006 WILEY-VCH Verlag GmbH & Co. KGaA, Weinheim

All rights reserved (including those of translation into other languages). No part of this book may be reproduced in any form – by photoprinting, microfilm, or any other means – nor transmitted or translated into a machine language without written permission from the publishers.
Registered names, trademarks, etc. used in this book, even when not specifically marked as such, are not to be considered unprotected by law.

Typesetting   Dörr + Schiller GmbH, Stuttgart
Printing   betz-druck GmbH, Darmstadt
Binding   Schäffer GmbH, Grünstadt
Cover Design   Grafik-Design Schulz, Fußgönheim

Printed in the Federal Republic of Germany
Printed on acid-free paper

ISBN-13:   978-3-527-31167-5
ISBN-10:   3-527-31167-X

Dedication

This book is dedicated to two outstanding scientists who have been my guide to the fascinating field of protein structure: Professor Robert Fletterick who introduced me to the beauty of protein structure and Professor Arthur Lesk who has taught me everything I know.

Foreword

The spontaneous folding of proteins to their native states is the point at which life makes the giant leap from the one-dimensional world of DNA and protein sequences to the three-dimensional world we inhabit. Proteins must therefore have an "algorithm" by which their sequences determine their structures. This discovery, by Anfinsen almost 50 years ago, has challenged scientists to reproduce this algorithm – or at least to find another, perhaps an artificial one – to predict protein structures from their sequences. This would truly unlock the immense stores of information contained in the many genome sequences now known, to reveal the evolution and development of biological function.

Claims of progress in protein structure prediction necessitate an objective approach to evaluating them. The CASP (Critical Assessment of Structure Prediction) programmes were devised as "blind tests" of developing methods. There is consensus that CASP has stimulated, as well as recorded, the recent advances in the field. CASP came just at the time when both the growth of sequence and structure data, and the improved power of algorithms and computers, joined to make progress possible.

Anna Tramontano has here set out the current state of the art. Combining the experience of a contributor with the skills of an expositor and teacher, she has organized and presented the field. In addition to her own contributions to the development of prediction methods, she has twice served as assessor at CASP meetings. Her book sets out structure prediction in the context of the biology of the problem – protein-folding itself, of course, and the background from studies of evolution of protein sequences, structures and functions that have provided the basis for many of the most successful methods of prediction.

Several factors contribute to the clarity of the book. The illustrations are well chosen. Web links are provided. At many points of the exposition questions and answers are interpolated, as if in a lecture a member of the audience raised a hand, was recognized, and the point explained.

Protein structure prediction is at the point of maturing from an esoteric specialty to a component of the standard tools of the molecular biologist. This book will catalyse that process. The book combines a presentation of the intellectual framework of the subject, with practical aspects: If I need a model, what method should I choose? How can I apply it? What can I expect from the result? How far can I trust it? To know how far to trust predictions, one must understand how the methods work. To contribute to the field, one must understand how to create new methods, areas where progress can be made, and how to evaluate one's own contribution (as well as those of others). This book is a good source for all these topics.

The book treats the mainline topics in protein structure prediction: Homology modelling, secondary structure prediction, fold recognition, and prediction of three-dimensional structures of proteins with novel folds. It also covers special cases, including membrane proteins and antibodies.

I and other readers must be grateful for this snapshot of state of the art at a crucial time. Of course methods will become more powerful – books like this sow the seeds of their own supersession, by fostering novel developments in the field. But this volume will remain a standard reference for the cusp of the wave.

Arthur Lesk
University Park, Pennsylvania, USA
November 2005

Table of Contents

Foreword *VII*

Preface *XII*

Acknowledgments *XV*

Introduction *XVI*

1 Sequence, Function, and Structure Relationships *1*
1.1 Introduction *1*
1.2 Protein Structure *4*
1.3 The Properties of Amino Acids *12*
1.4 Experimental Determination of Protein Structures *14*
1.5 The PDB Protein Structure Data Archive *20*
1.6 Classification of Protein Structures *22*
1.7 The Protein-folding Problem *24*
1.8 Inference of Function from Structure *27*
1.9 The Evolution of Protein Function *29*
1.10 The Evolution of Protein Structure *34*
1.11 Relationship Between Evolution of Sequence and Evolution of Structure *37*

2 Reliability of Methods for Prediction of Protein Structure *41*
2.1 Introduction *41*
2.2 Prediction of Secondary Structure *43*
2.3 Prediction of Tertiary Structure *46*
2.4 Benchmarking a Prediction Method *50*
2.5 Blind Automatic Assessments *51*
2.6 The CASP Experiments *51*

3	**Ab-initio Methods for Prediction of Protein Structures** 55
3.1	The Energy of a Protein Configuration 55
3.2	Interactions and Energies 55
3.3	Covalent Interactions 56
3.4	Electrostatic Interactions 58
3.5	Potential-energy Functions 62
3.6	Statistical-mechanics Potentials 62
3.7	Energy Minimization 65
3.8	Molecular Dynamics 66
3.9	Other Search Methods: Monte Carlo and Genetic Algorithms 67
3.10	Effectiveness of Ab-initio Methods for Folding a Protein 70

4	**Evolutionary-based Methods for Predicting Protein Structure: Comparative Modeling** 73
4.1	Introduction 73
4.2	Theoretical Basis of Comparative Modeling 75
4.3	Detection of Evolutionary Relationships from Sequences 77
4.4	The Needleman and Wunsch Algorithm 79
4.5	Substitution Matrices 81
4.6	Template(s) Identification Part I 84
4.7	The Problem of Domains 90
4.8	Alignment 91
4.9	Template(s) Identification Part II 96
4.10	Building the Main Chain of the Core 97
4.11	Building Structurally Divergent Regions 98
4.12	A Special Case: Immunoglobulins 102
4.13	Side-chains 106
4.14	Model Optimization 107
4.15	Other Approaches 108
4.16	Effectiveness of Comparative Modeling Methods 109

5	**Sequence-Structure Fitness Identification: Fold-recognition Methods** 117
5.1	The Theoretical Basis of Fold-recognition 117
5.2	Profile-based Methods for Fold-recognition 119
5.3	Threading Methods 121
5.4	Profile–Profile Methods 124
5.5	Construction and Optimization of the Model 124

6	**Methods Used to Predict New Folds: Fragment-based Methods** 127
6.1	Introduction 127
6.2	Fragment-based Methods 128
6.3	Splitting the Sequence into Fragments and Selecting Fragments from the Database 130
6.4	Generation of Structures 135

7	**Low-dimensionality Prediction: Secondary Structure and Contact Prediction** *137*
7.1	Introduction *137*
7.2	A Short History of Secondary structure Prediction Methods *140*
7.3	Automatic learning Methods *142*
7.3.1	Artificial Neural Networks *142*
7.3.2	Support Vector Machines *148*
7.4	Secondary structure Prediction Methods Based on Automatic Learning Techniques *150*
7.5	Prediction of Long-range Contacts *153*
8	**Membrane Proteins** *159*
8.1	Introduction *159*
8.2	Prediction of the Secondary Structure of Membrane Proteins *162*
8.3	The Hydrophobic Moment *165*
8.4	Prediction of the Topology of Membrane Proteins *166*
9	**Applications and Examples** *169*
9.1	Introduction *169*
9.2	Early Attempts *169*
9.3	The HIV Protease *171*
9.4	Leptin and Obesity *174*
9.5	The Envelope Glycoprotein of the Hepatitis C Virus *176*
9.6	HCV Protease *178*
9.7	Cyclic Nucleotide Gated Channels *181*
9.8	The Effectiveness of Models of Proteins in Drug Discovery *183*
9.9	The Effectiveness of Models of Proteins in X-ray Structure Solution *186*

Conclusions *188*

Glossary *190*

Index *201*

Preface

The enormous increase in data availability brought about by the genomic projects is paralleled by an equally unprecedented increase in expectations for new medical, pharmacological, environmental and biotechnological discoveries. Whether or not we will be able to meet, at least partially, these expectations it depends on how well we will be able to interpret the data and translate the mono-dimensional information encrypted in the genomes into a detailed understanding of its biological meaning at the phenotypic level.

The major components of living organisms are proteins, linear polymers of amino acids, whose specific sequence is dictated by the genes of the organism. The genes, linear polymers of nucleotides, are directly translated into the linear sequence of amino acids of the encoded protein through a practically universally conserved code. This conceptually simple process is indeed biochemically rather complex, it involves several cellular machineries and is subject to an intricate network of control mechanisms. Even the problem of identifying coding regions in eukaryotic genomes is not completely solved. Far more complex is the identification of the function of the encoded proteins and this will probably be the most challenging problem for the next generations of scientists.

Proteins mediate most of the functions of an organism, and all these functions are, in general, determined by the proteins' three-dimensional structure. Natural proteins spontaneously assume a unique three-dimensional structure. Although this is not a general property of polymers, it is shared, with rare exceptions, by all natural functional proteins. It is achieved by the interplay between molecular evolution and environment-driven selective pressure. Selective pressure acts on function, function requires structure, and therefore protein sequences are selected for being able to fold into a unique structure. This peculiarity of native protein sequences is the reason for their exceptional plasticity and versatility and leads to the exquisite specificity of these macromolecules and to very precise control of their activity through complex networks of interactions. If we could understand the relationship between protein sequence, structure, and function, we could make an effective use of the large body of genetic information available for many organisms, humans included, list all their functions, and study their interplay in shaping life.

Protein Structure Prediction. Edited by Anna Tramontano
Copyright © 2006 WILEY-VCH Verlag GmbH & Co. KGaA, Weinheim
ISBN: 3-527-31167-X

Protein sequence usually determines protein structure, as was established more than 50 years ago by Christian Anfinsen, the famous Nobel laureate American chemist who was the first to show that a protein can spontaneously refold to its native form and therefore that the information determining the three-dimensional structure of a protein resides in the chemistry of its amino acid sequence. Understanding the underlying rules is, however, a very challenging and as yet an unsolved problem, often referred to as the "holy grail" of biology. The problem has an enormous relevance in many fields, from medicine to biology, from biotechnology to pharmacology and therefore approximate methods for inferring the structure of a protein from its amino acid sequence are flourishing.

Such methods are an essential part of the cultural background and of the toolset of a biologist. There are several structure-prediction tools, easy to use and readily available via internet-based servers, and they are an important contribution of computational biology to the development of the life sciences. All available prediction methods have limitations, however, dictated by the hypotheses on which they rely, and this determines very precisely the field of application of the models produced. Before using any of them it is important to have a clear idea of the biological question for which the model should provide an answer and to verify whether the selected method is sufficiently accurate for the problem at hand. The latter depends upon several factors, and the aim of this book is to provide students and experimental researchers with a guide through the available methods, their underlying assumptions, their limitations, and their expected accuracy in different applications.

Before entering into the heart of the problem, we will discuss the relationship between sequence, structure, and function in proteins, to familiarize the reader with the features of a protein structure (Chapter 1) and with methods and initiatives aimed at evaluating the effectiveness and limitations of prediction methods, and their expected accuracy (Chapter 2).

In Chapter 3 we discuss the extent to which physics can help us compute the structure of a protein on the basis of the knowledge of its chemical structure. The possibility of predicting the structure of a protein from scratch using this approach is, unfortunately, out of our reach at present, and therefore we will explore alternative, knowledge based methods, in subsequent chapters. In Chapter 4 we will survey how a three-dimensional model of a protein can be built on the basis of its evolutionary relationship with a protein of known structure and how to detect the existence of such a relationship. Next, we will discuss how models of proteins can be built taking advantage of the observation that apparently unrelated proteins can share a similar overall structure (Chapter 5) or that, in any case, they share local structure similarities (Chapter 6). Methods which can be used to infer some structural properties of proteins, even if they do not provide us with their complete three-dimensional structure will be surveyed in Chapter 7. These methods are very useful for guiding the construction and assessing the reliability of atomic models obtained with other techniques.

Special treatment must be reserved to membrane proteins. Their peculiar properties, dictated by the environment in which they reside, enable different methods and strategies to be used for their prediction; these will be discussed in Chapter 8.

Finally, a description of a few successful examples of protein structure modeling will be discussed in the last chapter. The list could be much longer, the literature abounds with scientific reports in which structure prediction is used to guide experiments and to interpret data in the light of a model of the proteins of interest. The examples chosen here partially reflect the personal preference of the author and partially are meant to touch different aspects of the field of protein structure prediction.

The aim of the book is to convince the reader that protein modeling can be extremely useful, if it is performed carefully and with full understanding of the techniques, and that it should be application driven. There is no point in constructing a very approximate model of a protein and later to use it to derive detailed properties about catalytic mechanisms or interaction details; similarly, it is not wise to employ very sophisticated techniques to obtain a model of a protein that, for technical reasons, cannot be used for guiding experiments or for casting light on important biological questions.

Rome, November 2005 *Anna Tramontano*

Acknowledgments

I would like to express my profound gratitude to all my friends and colleagues who offered advice during the preparation of this book. I am deeply grateful to Domenico Cozzetto who took the time to read early drafts of this book and offer many valuable comments and criticisms.

Introduction

Proteins are the functioning molecules of living organisms; they play key roles in most physiological processes, for example metabolism, transport, immune response, signal transduction, and cell cycle. These linear polymers of amino acids perform this impressive variety of functions by assuming a well defined three-dimensional structure, which enables them to accomplish highly specific molecular functions. In these complex structures, amino acids that are far apart in the linear sequence can come close together in space and participate in the formation of highly specialized catalytic sites, ligand-binding pockets, or interfaces able to recognize, bind and transmit information to other macromolecules.

Genomic projects are providing us with the linear amino acid sequence of hundreds of thousands of proteins. If only we could learn how each and every one of these folds in three-dimensions we would have the complete part list of an organism and could face the challenge of understanding how these parts assemble in a cell. This is not only an intellectual challenge, it has enormous practical implications. Malfunctioning of proteins is the most common cause of endogenous diseases and the action of pathogens is usually mediated by their proteins. Most life-saving drugs act by interfering with the action of a faulty or foreign protein by keeping it from performing its function. Sometimes the drug competes directly with the substrate occupying the site where action occurs, sometimes it binds in a different location of the protein surface, causing a structural effect that modifies the geometry of the active site and makes the protein unable to perform its biochemical function. So far, most drugs have been discovered by trial-and-error, testing thousands of randomly selected compounds for their ability to interfere with the function of a protein or with a biological process. The active compounds are later modified to try and endow them with desirable properties such as higher specific activity, lack of toxicity, ability to reach the correct cellular compartment, favorable metabolic properties, ease of chemical synthesis, etc. The rate of success of this process is rather low, it is estimated that only one out of a thousand active molecules makes it to the pharmacy bench.

Our lack of a complete understanding of the complex interplay between different proteins has several implications for our welfare. The drugs we use might not be aimed at the best target, i.e. other proteins participating in the same biological process might be better targets. Drugs usually have undesirable side-effects that can only be assessed in clinical trials, because we cannot evaluate beforehand

Protein Structure Prediction. Edited by Anna Tramontano
Copyright © 2006 WILEY-VCH Verlag GmbH & Co. KGaA, Weinheim
ISBN: 3-527-31167-X

whether they bind and interfere with molecules involved in different processes. Differences between the genetic background of individuals might cause the protein of a specific individual to be less sensitive to the effect of a drug or cause the drug to have more serious side effects. All this makes the drug-development process very time consuming and far from optimum. In an ideal situation, if we had a deep structural understanding of our complete parts list, we could screen drugs in a virtual in-silico system and detect a substantial proportion of these problems beforehand, increasing substantially the effectiveness of medical approaches. To achieve this goal we need to know the structure of the proteins involved at a very accurate level of detail.

Experimental methods can provide us with knowledge of the precise arrangement of every atom of a protein. The most effective of these are X-ray crystallography and NMR spectroscopy. X-ray crystallography, however, requires the protein or the protein complex under study to form a reasonably well ordered crystal, a feature that is not universally shared by proteins. NMR spectroscopy needs proteins to be soluble and there is a limit to the size of protein that can be studied. The structural information on some proteins can be very difficult to obtain experimentally. For example, proteins embedded in hydrophobic biological membranes are both difficult to crystallize and insoluble in polar solvents. Yet, these proteins play fundamental roles in biology because they are involved in the process of communication between cells and in the uptake of external molecules. Both X-ray crystallography and NMR are time-consuming techniques and we cannot hope to use them to solve the structures of all the proteins of the universe in the near future. As will be discussed, we know that the structure of a protein depends solely on its amino acid sequence, therefore we would like to develop theoretical approaches to infer (or predict, as we say) the structure of a protein from its amino acid sequence.

For decades, the problem of deciphering the code that relates the amino acid sequence of a protein and its native three-dimensional structure has been the subject of innumerable investigations and, despite the many frustrations caused by its elusiveness, interest in the problem is not fading away. On the contrary. What stands in our way, notwithstanding all these efforts, is the complexity of protein structures. In a three-dimensional protein structure, thousands of atoms are held together by weak forces and give rise to a conformation which is only marginally stable. The consequence of this, as we will discuss, is that it is very unlikely that we can use our understanding of the laws of physics to compute the native functional structure of a protein in the foreseeable future. We have at our disposal, however, the experimentally solved structures of a reasonable number of proteins, currently a few thousand. They represent solved instances of our problem and we can hope to extract heuristic rules from their analysis. This is the topic of this book – it describes what we can learn from the analysis of known protein structures and how this knowledge can be used to attempt the prediction of unknown protein structures. As we will see, this approach has led to several methods for protein structure prediction, some reaching a respectable level of accuracy and reliability. Usually, however, the accuracy of a model is not comparable with that achievable by experimental methods, and it is therefore important to understand the limitations of the methods.

Now that the sequences of many proteins are available to us and computers are becoming increasingly powerful, it is possible to run high throughput modeling experiments and predict the structures of thousands of proteins automatically. In other, more difficult instances, models must be built manually, taking into account other data, especially experimental, and each step of the process must be accurately analyzed for the possible presence of errors. In both instances, but even more in the latter, it is important to ask a few questions before approaching the problem of predicting the structure of a protein:

- What will be the application of the model?
- Is the expected accuracy of the obtainable model appropriate for the task?
- Can we devise experiments to verify the correctness of our model before we use it to infer the properties of the target protein?

Models of proteins can, undoubtedly, be very useful for several applications and we will describe some examples in which they have been instrumental in achieving better understanding of a difficult biological problem, but it should always be kept in mind that there are limits to their accuracy. Admittedly, experimental determinations of protein structures are also affected by errors, as are all experimental data, but at present, except for a handful of cases, the accuracy of an experimental structure is far superior to that achievable with modeling methods.

The situation is, therefore, that we have a limited number of protein structures determined with high accuracy and less detailed information derived by modeling on a much larger number of proteins. There is no doubt that the most efficient strategy is for the two fields to complement each other in exploring the protein structure space. We should optimize efforts and solve experimentally the structure of proteins selected in a way that increases the chances that modeling methods can use them to produce reasonably accurate models of many others. Worldwide initiatives, denoted structural genomics projects, are indeed following this idea and are producing the structures of a large number of proteins selected with the aim of providing representative examples of the protein structural space. The task of protein structure modeling is to make the best use of the available experimental protein structure information to infer the structures of as many of the remaining proteins as possible.

Traditionally, protein structure prediction methods have been categorized according to the nature of the relationship between the protein to be modeled and the available proteins of known structure. Comparative or homology modeling is based on the observation that evolutionarily related proteins preserve their structural features during evolution and is the method of choice when a clear evolutionary relationship can be detected between the target protein and one or more proteins of known structure. Even proteins with no clear evolutionary relationship often share a similar structure and methods devoted to detecting these and exploiting this information to model a protein are called "fold-recognition methods". Finally, in all other instances, the observation that some structural features are shared between proteins not falling into any of the two previous categories, gave rise to methods known as "fragment-based methods". The boundaries between the three methodologies are, as we will see, becoming less defined, and methods are cross-fertilizing each other in several ways. Nevertheless, here we will still follow the traditional subdivision between methods, because we believe that it makes it easier to navigate through the many aspects of this fascinating problem.

1
Sequence, Function, and Structure Relationships

1.1
Introduction

Life is the ability to metabolize nutrients, respond to external stimuli, grow, reproduce, and, most importantly, evolve. Most of these functions are performed by proteins, organic macromolecules involved in nearly every aspect of the biochemistry and physiology of living organisms. They can serve as structural material, catalysts, adaptors, hormones, transporters, regulators. Chemically, proteins are linear polymers of amino acids, a class of organic compounds in which a carbon atom (called Cα) is bound to an amino group (–NH$_2$), a carboxyl group (–COOH), a hydrogen atom (H), and an organic side group (called R). The physical and chemical properties unique to each amino acid result from the properties of the R group (Figure 1.1).

In a protein, the amino group of one amino acid is linked to the carboxyl group of its neighbor, forming a peptide (C–N) bond. There are two resonance forms of the peptide bond (i.e. two forms that differ only in the placement of electrons), as illustrated in Figure 1.2. Atoms involved in single bonds share one pair of electrons whereas two pairs are shared in double bonds. The latter are planar, i.e. not free to rotate. The resonance between the two forms shown in Figure 1.2 makes the peptide bond intermediate between a single and a double bond and, as a consequence, all peptide bonds in protein structures are found to be almost planar – the Cα, C, and O atoms of one amino acid and the N, H, and Cα of the next lie on the same plane. Although this rigidity of the peptide bond reduces the degrees of freedom of the polypeptide, the dihedral angles around the N–Cα and the Cα–C bonds are free to vary and their values determine the conformation of the amino acid chain.

Proteins assume a three-dimensional shape which is usually responsible for their function. The consequence of this tight link between structure and function and of the evolutionary pressure to preserve function has a very important effect – in contrast with ordinary polymers (e.g. polypeptides with random sequences) that typically form amorphous globules, proteins usually fold to unique structures. In other words they spontaneously assume a unique three-dimensional structure specified, as we will see, by their amino acid sequence. For example, enzymes accelerate chemical reactions by stabilizing their high energy intermediate and this

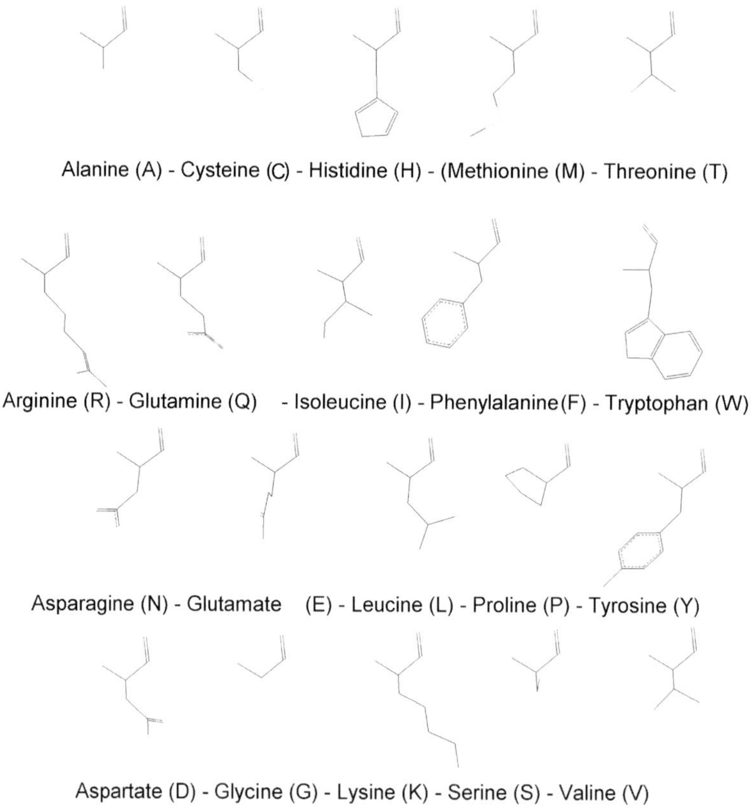

Alanine (A) - Cysteine (C) - Histidine (H) - (Methionine (M) - Threonine (T)

Arginine (R) - Glutamine (Q) - Isoleucine (I) - Phenylalanine (F) - Tryptophan (W)

Asparagine (N) - Glutamate (E) - Leucine (L) - Proline (P) - Tyrosine (Y)

Aspartate (D) - Glycine (G) - Lysine (K) - Serine (S) - Valine (V)

Figure 1.1 The twenty naturally occurring amino acids.

is achieved by correct relative positioning of appropriate chemical groups. Our body contains many proteins that catalyze the hydrolysis of peptide bonds in proteins (the inverse of the polymerizing reaction used to build proteins) to provide the body with a steady supply of amino acids. The substrates of these reactions are either proteins from the diet or "used" proteins inside the body. Digestion begins in the stomach where the acidic environment unfolds, i.e. destructures, the proteins and an enzyme called pepsin (Figure 1.3) starts chopping the proteins into pieces. Later, in the intestines, several protein-cutting enzymes, for example trypsin (also shown in Figure 1.3), cut the protein chains into shorter pieces. In subsequent steps, other enzymes reduce these shorter pieces to single amino acids.

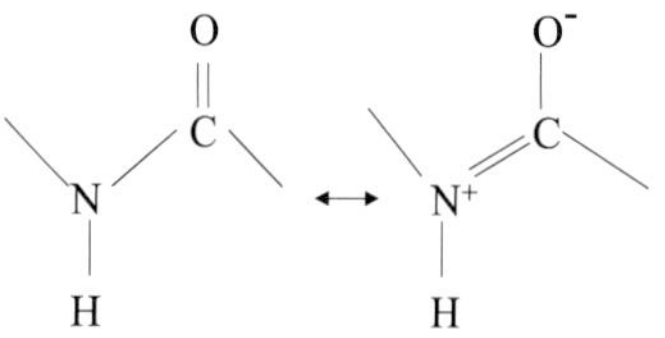

Figure 1.2 The two resonance structures of the peptide bond. Because of delocalization of the electrons, the C–N bond has the character of a partial double bond and this limits its freedom of rotation.

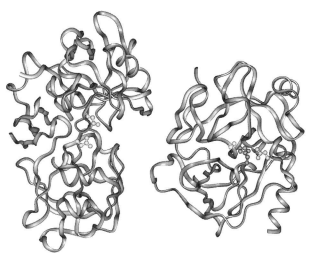

```
IGDEPLENYL  DTEYFGTIGI  GTPAQDFTVI  FDTGSSNLWV      IVGGYTCGAN  TVPYQVSLNS  GYHFCGGSLI  NSQWVVSAAH
PSVYCSSLAC  SDHNQFNPDD  SSTFEATSQE  LSITYGTGSM      CYKSGIQVRL  GEDNINVVEG  NEQFISASKS  IVHPSYNSNT
TGILGYDTVQ  VGGISDTNQI  FGLSETEPGS  FLYYAPFDGI      LNNDIMLIKL  KSAASLNSRV  ASISLPTSCA  SAGTQCLISG
LGLAYPSISA  SGATPVFDNL  WDQGLVSQDL  FSVYLSSNDD      WGNTKSSGTS  YPDVLKCLKA  PILSDSSCKS  AYPGQITSNM
SGSVVLLGGI  DSSYYTGSLN  WVPVSVEGYW  QITLDSITMD      FCAGYLEGGK  DSCQGDSGGP  VVCSGKLQGI  VSWGSGCAQK
GETIACSGGC  QAIVDTGTSL  LTGPTSAIAN  IQSDIGASEN      NKPGVYTKVC  NYVSWIKQTI  ASN
SDGEMVISCS  SIASLPDIVF  TINGVQYPLS  PSAYILQDDD
SCTSGFEGMD  VPTSSGELWI  LGDVFIRQYY  TVFDRANNKV
GLAPVA
```

Figure 1.3 The three-dimensional structures of pepsin (left, PDB code: 1PSN) and trypsin (right, PDB code: 3TPI). These two enzymes cleave peptide bonds with a different mechanism. The first uses two aspartic acids, the second a triad formed by a histidine, a serine, and an aspartic acid. Their amino acid sequence is shown at the bottom of the figure in a one-letter code. Note that, for both enzymes, the amino acids forming the active site (underlined) are distant in the linear sequence and are brought together by the three-dimensional structure of the enzymes.

Both pepsin and trypsin belong to a class of enzymes called proteases, the first performs its job by taking advantage of the presence, in a cleft of the protein structure of two residues of aspartic acid; in the second catalysis is achieved by cooperation of three amino acids, a serine, a histidine, and an aspartic acid. Amino acids near the active site are responsible for recognition and correct positioning of the substrate. These functional amino acids, far apart in the linear amino acid sequence (Figure 1.3), are brought together in exactly the right position by the protein three-dimensional structure.

Similarly, recognition of foreign molecules is mediated by several proteins of the immune system, the most popular being antibodies. Antibodies bind other molecules, called antigens, by means of an exposed molecular surface complementary to the surface of the antigen, which can be a protein, a nucleic acid, a polysaccharide, etc. The binding surface is formed by amino acids from different parts of the molecule (Figure 1.4).

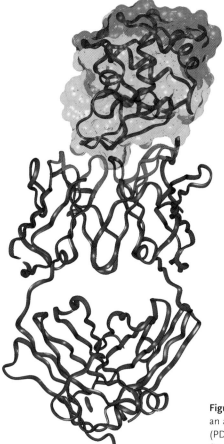

Figure 1.4 The three-dimensional structure of an antibody bound to its cognate molecule (PDB code: 3HFL). Note that the binding region (in red) is formed by amino acids from different regions of the linear sequence.

1.2
Protein Structure

Most readers will already be familiar with the basic concepts of protein structure; we will, nevertheless, review here some important aspects of this subject. The sequence of amino acids, i.e. of the R-groups, along the chain is called the primary structure. Secondary structure refers to local folding of the polypeptide chain. Tertiary structure is the arrangement of secondary structure elements in three dimensions and quaternary structure describes the arrangement of a protein's subunits. As we have already mentioned, the peptide bond is planar and the dihedral angle it defines is almost always 180°. Occasionally the peptide bond can be in the cis conformation, i.e. very close to 0°.

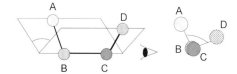

Figure 1.5 A dihedral angle between four points A, B, C, and D is the angle between two planes defined by the points A, B, C and B, C, D, respectively.

Question: What is a dihedral angle?

»The dihedral angle is the angle between two planes. In practice, if you have four connected atoms and you want to measure the dihedral angle around the central bond, you orient the system in such a way that the two central atoms are superimposed and measure the resulting angle between the first and last atom (Figure 1.5).«

The simplest arrangement of amino acids that results in a regular structure is the alpha helix, a right-handed spiral conformation. The structure repeats itself every 5.4 Å along the helix axis. Alpha helices have 3.6 amino acid residues per turn and

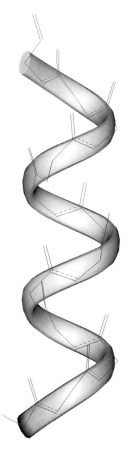

Figure 1.6 An alpha helix. Each backbone oxygen atom is hydrogen-bonded to the nitrogen of a residue four positions down the chain.

the separation of the residues along the helix axis is 5.4/3.6 or 1.5 Å, i.e. the alpha helix has a rise per residue of 1.5 Å. Every main-chain C=O group forms a hydrogen bond with the NH group of the peptide bond four residues away (Figure 1.6).

Let us recall that a hydrogen bond is an intermolecular interaction formed between a hydrogen atom covalently bonded to an electronegative atom (for example oxygen or nitrogen) and a second electronegative atom that serves as the hydrogen-bond acceptor. The donor atom, the hydrogen, and the acceptor atom are usually co-linear. The alpha helix has 3.6 residues per turn and thirteen atoms enclosed in the ring formed by the hydrogen bond, it can also be called a 3.6(13) helix. Another type of helix is observed in protein structures, although much more rarely; this is the 3(10) helix. This arrangement contains three residues per turn and ten atoms in the ring formed by the hydrogen bond. In alpha helices, the peptide planes are approximately parallel with the helix axis, all C=O groups point in one direction, and all N–H groups in the opposite direction. Because of the partial charge on these groups, negative for CO and positive for NH, there is a resulting dipole moment in the helix. Side-chains point outward and pack against each other. The dipoles of a 3(10) helix are less well aligned and the side-chain packing less favorable, therefore it is usually less stable. Typically, in alpha helices the angles around the N–Cα and Cα–C bonds, called φ and ψ angles, are approximately –60° and –50°, respectively.

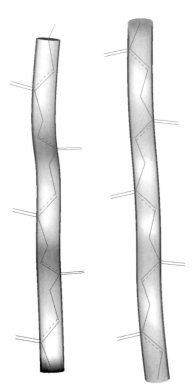

Figure 1.7 Two beta strands forming an antiparallel beta sheet. Oxygen and nitrogen atoms of different strands are hydrogen bonded to each other.

Another secondary structure element commonly observed in proteins is the beta sheet, an arrangement of two or more polypeptide chains (beta strands) linked in a regular manner by hydrogen bonds between the main chain C=O and N–H groups. The R groups of neighboring residues in a beta strand point in opposite directions forming a layered structure (Figure 1.7). The strands linked by the hydrogen bonds in a beta sheet can all run in the same direction (parallel sheet) or in opposite directions (antiparallel sheet). Beta sheets can be mixed, including both parallel and antiparallel pairs of strands. Most beta sheets found in proteins are twisted – each residue rotates by approximately 30° in a right-handed sense relative to the previous one.

The plot shown in Figure 1.8 is called a Ramachandran plot. This can be obtained by considering atoms as hard spheres and recording which pairs of φ and ψ angles do not cause the atoms of a dipeptide to collide. Allowed pairs of values are represented by dark regions in the plot whereas sterically disallowed regions are left white. The lighter areas are obtained by using slightly smaller radii of the spheres, i.e. by allowing atoms to come a bit closer together. Disallowed regions usually involve steric hindrance between the first carbon atom of the side-chain, the Cβ, and main-chain atoms. As we will see, the amino acid glycine has no side-chain and can adopt φ and ψ values that are unfavorable for other amino acids.

> Question: Do we observe amino acids with dihedral angles in disallowed regions of the Ramachandran plot in experimental protein structures?
>
> »Yes, we do. Even in very well refined crystal structures of proteins at high resolution, some φ and ψ angles fall into disallowed regions. The reader should keep in mind that the reason some combinations of angles are rarely observed is because they are energetically disfavored, not mathematically impossible. The loss of energy because of an unfavorable dihedral angles combination can be compensated by other interactions within the protein.«

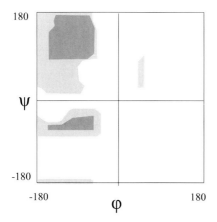

Figure 1.8 A Ramachandran plot is a graph reporting the values of phi and psi angles in protein structures. Darker areas indicate favorable combinations of angles, lighter gray areas are less favored, but still possible.

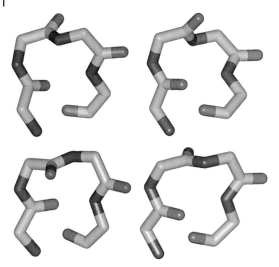

Figure 1.9 The four types of beta turn described in Table 1.1, types I and I' are shown on the top, types II and II' on the bottom.

Regions without repetitive structure connecting secondary structure elements in a protein structure are called loops. The amino acid chain can reverse its direction by forming a reverse turn characterized by a hydrogen bond between one main chain carbonyl oxygen and the N–H group 3 residues along the chain (Figure 1.9). When such a secondary structure element occurs between two anti-parallel adjacent beta strands in a beta sheet is called a beta hairpin. Reverse turns are classified on the basis of the ϕ and ψ angles of the two residues in their central positions as shown in Table 1.1. Note that some turns require that one of their amino acids has ϕ and ψ angles falling in disfavored regions of the Ramachandran plot.

Table 1.1 Turns are regions of the protein chain that enable the chain to invert its direction. The ϕ and ψ angles of some commonly occurring turns are listed.

Turn type	ϕ_1	ψ_1	ϕ_2	ψ_2
I	−60	−30	−90	0
I'	60	30	90	0
II	−60	120	80	0
II'	60	−120	−80	0

Proteins can be formed from only alpha helical or from only beta sheet elements, or from both; the association of these elements within a single protein chain is called tertiary structure. Certain arrangements of two or three consecutive secondary structures (alpha helices or beta strands), are present in many different protein

1.2 Protein Structure

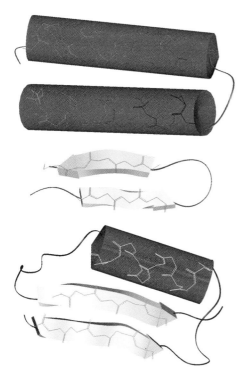

Figure 1.10 Supersecondary structures: alpha–loop–alpha, beta hairpin, and beta-alpha-beta unit.

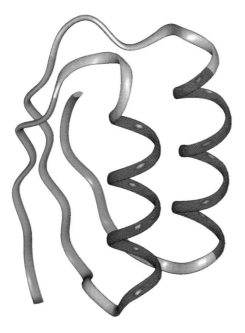

Figure 1.11 The Rossmann fold. The figure shows a region of the succinyl-Coa synthetase enzyme from the bacterium *Escherichia coli* (PDB code: 2SCU).

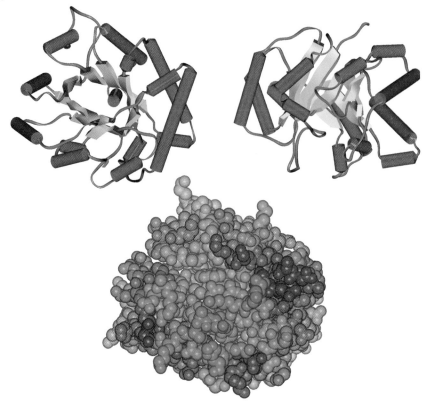

Figure 1.12 A TIM barrel (PDB code: 8TIM). The structure at the top left is the same as that at the top right rotated by 90° around an horizontal axis. On the bottom the structure is shown with all its non-hydrogen atoms. Atoms in green belong to the central beta barrel, atoms in red to the surrounding helices.

structures, even with completely different sequences; these are called supersecondary structures. They include the alpha–alpha unit (two antiparallel alpha helices joined by a turn); the beta–beta unit (two antiparallel strands connected by a hairpin); and the beta–alpha–beta unit (two parallel strands, separated by an alpha helix antiparallel to them (Figure 1.10). Sometimes the term "motif" is used to describe these supersecondary structures. Supersecondary structures are not necessarily present in a protein structure, however, which can be formed from several alpha helices or beta strands without containing any of the supersecondary structures described above. On the other hand, some combinations of the supersecondary structural motifs are observed relatively often in proteins. A very commonly found arrangement of helices is the four-helix bundle (two alpha–alpha units connected by a loop). Another common motif is the beta–alpha–beta–alpha–beta unit, called the Rossman fold (Figure 1.11). These arrangements are often called domains or folds. Some folds can be very large and complex and can be formed from several supersecondary structures. One example is the TIM barrel fold; this is shared by many enzymes and formed from several beta–alpha–beta units (Figure 1.12).

Another layer of organization of protein structure is the domain level. The definition of a domain is rather vague. Some confusion also arises because the term is often also used in the context of the amino acid sequence, rather than of its three-dimensional structure. In general a domain can be defined as a portion of the polypeptide chain that folds into a compact semi-independent unit. Domains can be seen as "lobes" of the protein structure that seem to have more interaction between themselves than with the rest of the chain (Figure 1.13). Several proteins are formed from many repeated copies of one or a few domains; such proteins are called mosaic proteins and the domains are often referred to as "modules". A domain can be formed by only (or almost only) alpha helices or beta sheets, or by their combination. In the latter case the helices and strands can be packed against each other in the beta–alpha–beta supersecondary arrangement (alpha/beta domains) or separated in the structure (alpha + beta domain).

Finally, we talk about architecture of a protein when we consider the orientations of secondary structures and their packing pattern, irrespective of their sequential order, and we talk of protein topology when we also take into account the nature of the connecting loops and, therefore, the order in which the secondary structure elements occur in the amino acid sequence.

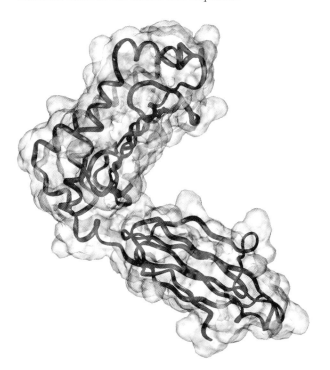

Figure 1.13 A two-domain protein chain (PDB code: 1HSA).

1.3
The Properties of Amino Acids

There are twenty naturally occurring amino acids. They can play different roles and it is important to survey their properties to be able to analyze and ultimately attempt to predict the structure and function of a protein.

The smallest amino acid is glycine, the side-chain of which is just a hydrogen atom. The lack of a side-chain makes this amino acid very flexible. We have already mentioned that this amino acid can assume "unusual" ϕ and ψ angles. We also saw that the structural requirements of turns often need an amino acid in this conformation and indeed these positions are often occupied by glycines. The observation that a glycine is always present in a given position in a family of evolutionarily related proteins often points to the presence of a tight turn in the region. The flexibility of glycine also implies that the loss of entropy associated with restricting its conformation in a protein structure is higher than for other amino acids, and the absence of a side-chain makes it less likely for this amino acid to establish favorable interactions with surrounding amino acids. Glycines are, therefore, rarely observed in both alpha helices or beta sheets.

The next amino acid, in order of size, is alanine. Here the side-chain is a CH_3 group. It is a small hydrophobic amino acid, without any reactive group, and rarely involved in catalytic function. Its small non polar surface and its hydrophobic character suggest, however, that this amino acid can be exposed to solvent, without large loss of entropy, and can also establish favorable hydrophobic interactions with other hydrophobic surfaces. In other words, it is an ideal amino acid for participating in interacting surfaces between proteins that associate transiently.

Cysteine is a small hydrophobic amino acid that can form disulfide bridges, i.e. covalent bonds arising as a result of the oxidation of the sulfhydryl (SH) group of the side-chains of two cysteine units when they are in the correct geometric orientation (Figure 1.14). Disulfide bridges enable different parts of the chain to be covalently bound. Because the intracellular environment is reducing, disulfide bridges are only observed in extracellular proteins. Cysteine can also coordinate metals and its SH group is rather reactive. In some viral proteases it takes the role of serine in serine protease active sites we have already described.

Serine is a small polar amino acid found both in the interior of proteins and on their surfaces. It is sometimes found within tight turns, because of its small size and its ability to form a hydrogen bond with the protein backbone. It is often

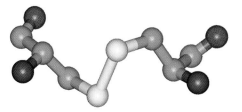

Figure 1.14 A disulfide bond. The yellow atoms are sulfur atoms.

observed in active sites, where it can act as a nucleophile as already mentioned for serine proteases. Another important property of this amino acid is that it is a substrate for phosphorylation – enzymes called protein kinases can attach a phosphate group to its side-chain. This plays important roles in many cellular processes and in signal transduction.

Another relatively small amino acid, rather similar to serine, is threonine. This amino acid can also be part of active sites and can be phosphorylated. An important difference, though, is that threonine is "beta branched", i.e. it has a substituent on its beta carbon and this makes it less flexible and less easy to accommodate in alpha helices. Beta-branched amino acids are indeed more often found in beta sheets.

Asparagine and glutamine are polar amino acids that generally occur on the surface of proteins, exposed to an aqueous environment, and frequently involved in active sites. Asparagine, for example, is found as a replacement for aspartate in some serine proteases. One peculiar property of asparagine is that it is often found in the left-handed conformation (positive ϕ and ψ angles) and can therefore play a role similar to that of glycine in turns. This is possibly because of its ability to form hydrogen bonds with the backbone.

Proline is unique because it is an imino acid rather than an amino acid. This simply means that its side-chain is connected to the protein backbone twice, forming a five-membered nitrogen-containing ring. This restricts its conformational flexibility and makes it unable to form one of the two main-chain hydrogen bonds that other amino acids can form in secondary structure elements; it is, therefore, often found in turns in protein structures. When it is in an alpha helix, it induces a kink in the axis of the helix. It is not a very reactive amino acid, but plays an important role in molecular recognition – peptides containing prolines are recognized by modules that are part of many signaling cascades. Proline can be found in the cis conformation (i.e. with the angle around the peptide bond close to 0° rather than 180°). The main chain nitrogen atoms of the other amino acids are bound to a hydrogen and a carbon atom whereas the situation in proline is more symmetrical with the atom bound to two carbon atoms. This means that the energy difference between the *cis* and *trans* conformations is smaller for this amino acid.

Leucine, valine, and isoleucine are hydrophobic amino acids, very rarely involved in active sites. The last two are beta branched and therefore often found in beta sheets and rarely in alpha *helices*.

Aspartate and glutamate are negatively charged amino acids, generally found on the surface of proteins. When buried, they are involved in salt bridges, i.e. they form strong hydrogen bonds with positively charged amino acids. They are frequently found in protein active sites and can bind cations such as zinc.

Lysine and Arginine are positively charged and can have an important role in structure. The first part of their side-chain is hydrophobic, so these amino acids can be found with part of the side-chain buried, and the charged portion exposed to solvent. Like aspartate and glutamate, lysine and arginine can form salt bridges and occur quite frequently in protein active or binding sites. They are, furthermore, often involved in binding negatively charged phosphates and in the interacting surfaces of DNA- or RNA-binding proteins.

At physiological pH, histidine can act as both a base or an acid, i.e. it can both donate and accept protons. This is an important property that makes it an ideal residue for protein functional centers such as the serine protease catalytic triad. Histidine can, furthermore, bind metals (e.g. zinc). This property is often exploited to simplify purification of proteins cloned and expressed in heterologous systems. The addition of a tail of histidines to the protein of interest confers on the protein the ability to chelate metals and this engineered property can be exploited for purifying the protein.

Methionine has a long and flexible hydrophobic side-chain. It is usually found in the interior of proteins. Like cysteine, it contains a sulfur atom, but in methionine the sulfur atom is bound to a methyl group, which makes it much less reactive.

Phenylalanine, tryptophan, and tyrosine are aromatic amino acids. The term "aromatic" was used by chemists to describe molecules with peculiar odors long before their chemical properties were understood. In chemistry, a molecule is called aromatic if it has a planar ring with $4n + 2$ π-electrons where n is a non-negative integer (Hückel's Rule). In practice, these molecules, the prototype of which is benzene, have a continuous orbital overlap that gives them special optical properties. For example, tryptophan absorbs light at 280 nm and this property is routinely used to measure the concentration of proteins in a solution (assuming the protein contains at least one tryptophan). Also, if an aromatic residue is held rigidly in space in an asymmetric environment, it absorbs left-handed and right-handed polarized light differently. This effect, which can be measured by circular dichroism spectroscopy, is therefore sensitive to overall three-dimensional structure and can be used to monitor the conformational state of a protein. Another important property of amino acids with aromatic side-chains is that they can interact favorably with each other. The face of an aromatic molecule is electron-rich while the hydrogen atoms around the edge are electron-poor. This implies that off-set face-to-face and edge-to-face interactions between aromatic rings have both hydrophobic and electrostatic components. Tyrosine is also a substrate for phosphorylation, similarly to serine and threonine, although the enzymes responsible for phosphorylation of tyrosine are different from those that phosphorylate serine and threonine.

1.4
Experimental Determination of Protein Structures

Two experimental techniques are used to determine the three-dimensional structure of macromolecules at the atomic level – X-ray crystallography and nuclear magnetic resonance (NMR) spectroscopy. Although it is beyond the scope of this book to describe the details of these techniques, which are rather complex both theoretically and experimentally, it is important to have some basic understanding of their results, because, as we will see, most methods for prediction of protein structure are based on existing structural data.

X-ray crystallography is based on the fact that an ordered ensemble of molecules arranged in a crystal lattice diffracts X-rays (the wavelengths of which are of the order of interatomic distances) when hit by an incident beam. The X-rays are dispersed by the electrons in the molecule and interfere with each other giving rise to a pattern of maxima and minima of diffracted intensities which depends upon the position of the electrons (and hence of the atoms) in the ordered molecules in the crystal. The electron density of the protein, i.e. the positions of the protein atoms, determines the diffraction pattern of the crystal, that is the magnitudes and phases of the X-ray diffraction waves, and vice versa, through a Fourier transform function. In practice:

$$\varrho(x,y,z) = \frac{1}{V}\sum_{hkl}\vec{F}_{hkl} = \frac{1}{V}\sum_{h}\sum_{k}\sum_{l} F(h,k,l)e^{-2\pi i(hx+ky+lz)}$$

where $\varrho(x,y,z)$ is the electron density at position (x,y,z), V is the volume, $\vec{F}(h,k,l)$ is the vector describing the diffracted waves in terms of their amplitudes $F(h,k,l)$ and phases (the exponential complex term). The electron density at each point depends on the sum of all of the amplitudes and phases of each reflection. If we could measure the amplitude and phase of the diffracted waves, we could relatively easily compute the exact relative location of each atom in the diffracting molecules. Unfortunately the phase of the diffracted waves cannot be measured and, therefore, we must use "tricks" to guess their approximate value and reconstruct the image of the diffracting molecule.

In effect, three methods are used to estimate the phases. Direct methods consist in using all possible values for the phases in the Fourier transform equation until an interpretable electron density is found; this is feasible for small molecules only. Interference-based methods can make use of multiple isomorphous replacement or anomalous scattering techniques. The first derives the phase by comparing the diffraction pattern of a protein crystal with that of crystals identical to the original one but for the presence of "heavy" atoms (i.e. atoms with many electrons and, therefore, very strong diffracting power) in specific positions of the molecules. The "anomalous scattering" technique instead derives initial phases by measuring diffraction data at several different wavelengths near the absorption edge of a heavy-atom. Finally, if we have a reasonable model for the molecule in the crystal, we can resort to the "molecular replacement" technique which computes approximated phases for the molecule in the crystal on the basis of the position of the atoms in the model. The availability of a high-quality three-dimensional model for a protein can therefore also be instrumental in obtaining its experimentally determined structure.

Given the diffracted intensities of a protein crystal and a set of "good" estimated phases, we can calculate the electron density that formed the observed pattern and position the atoms of the protein in the computed electron density (Figure 1.15). Important aspects of the whole procedure are that the protein under examination forms well ordered, well diffracting crystals and that the phase estimation procedure is successful in generating an interpretable electron-density map.

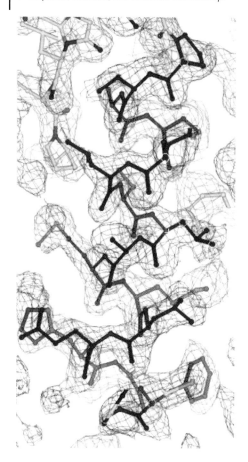

Figure 1.15 An electron-density map derived from an X-ray crystallography experiment. The atoms can be positioned in the map as shown in the figure, revealing an alpha helical structure.

Question: Is the quality of an X-ray determination of a protein structure comparable to that for small organic molecules?

»The quality of the structural data that can be obtained by protein crystallography is nowhere near the accuracy with which crystal structures of small molecules can be determined. This is because proteins can assume many different, although closely related, conformations and this limits the order of the molecules within the crystal. Also, protein crystals are usually only about half protein – the other half is occupied by solvent molecules. As we will see, the accuracy of small molecule crystallography can be used to derive parameters useful in modeling procedures.«

Just as in every experiment, in protein crystallography also the quality of the results improves with the ratio of the amount of data collected (the diffraction intensities) and the number of properties estimated (the positions of the atoms). In crystallography, the inverse of this ratio is expressed by the term "resolution" which is expressed in Angstroms (1 Angstrom = 10^{-10} m). The lower is the resolution the better is the quality of the structure. A resolution of approximately 3Å enables secondary structural elements and the direction of the polypeptide chain to be clearly identified in the electron density map; with a resolution of 2.5Å the side-chains can be built into the map with reasonable precision.

Hydrogen atoms do not diffract very well, because they only have one electron, and they are usually not detectable by X-ray crystallography unless the resolution is really very good, approximately 1.0Å. This implies that, given a crystal structure with good but not exceptional resolution, we can only deduce the presence of hydrogen bonds by the position of the donor and acceptor atoms.

After reconstructing the structure, we can back compute the expected diffraction pattern and compare it with that observed. The R factor indicates how much the two patterns (theoretical and experimental) differ and is expressed as a percentage. This factor is linked to the resolution. As a rule of thumb, a good structure should have an R factor lower than the resolution divided by 10 (i.e. $\leq 30\%$ for a 3.0Å resolution structure, $\leq 20\%$ for a 2.0Å structure, etc). To avoid any bias, it is more appropriate to compare the expected data with data set aside and not used to reconstruct the structure. In this case the term is called "Rfree". For a correctly reconstructed structure, one expects the ratio $R/Rfree$ to be $>80\%$.

Of course atoms in a crystal also have thermal motion. We can estimate the extent of their motion by looking at their electron density and, indeed, crystallography assigns a value that describes the extent to which the electron density is spread out to each atom. This value, called the "temperature factor" or "Debye–Waller factor" or B factor is given by:

$$B = 8\pi^2 U^2$$

where U is the mean displacement of the atom (in Å), so high B factors indicate greater uncertainty about the actual atom position. For example, for $B = 20$ Å^2:

$$U = \frac{1}{\pi}\sqrt{\frac{B}{8}}\text{Å} = \frac{1}{\pi}\sqrt{\frac{20}{8}}\text{Å} \cong \frac{1}{3.14}\sqrt{2.5}\text{Å} \cong 0.5\text{Å}$$

and the uncertainty about the position of the atoms is 0.5 Å. Values of 60 or greater may imply disorder (for example, free movement of a side-chain or alternative side-chain conformations). As expected, atoms with higher B factors are often located on the surface of a protein whereas the positions of the atoms in the internal well packed core of the protein are less uncertain (Figure 1.16). Finally, the occupancy value for an atom represents the fraction of expected electron density that was actually observed in the experiment.

Nuclear magnetic resonance (NMR) is another very useful technique for determining the structure of macromolecules. This technique is based on the observation that several nuclei (e.g. H, ^{13}C, ^{15}N) have an intrinsic magnetic moment. If we place a concentrated homogeneous solution of a protein (or nucleic acid) inside a

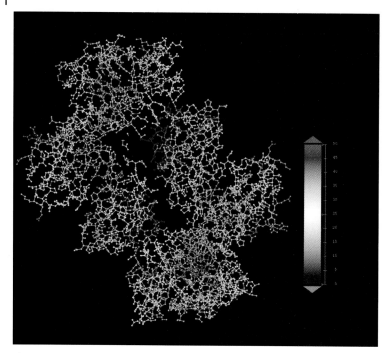

Figure 1.16 A protein structure colored according to the B-factor of its atoms. The color scheme is such that atoms with high B-factors are red and those with low B-factors blue.

very powerful magnetic field, the spin of the nuclei will become oriented in the direction of the external field. By applying radio-frequency magnetic fields to the sample, we can measure the energy absorbed at the frequency corresponding to the jump between two allowed spin orientations. Each atom has a characteristic resonance which depends on its structure, but it is also affected by the surrounding atoms. These subtle absorbance differences between the same atom in different environments make it possible to identify which resonance corresponds to which of the protein atoms.

If two atoms are close in space, magnetic interactions between their spins can be measured. The intensity of the interaction decays rapidly with the distance between the atoms (it is a function of r^{-6}, where r is the distance). This effect can be exploited to map short distances between interacting atoms. The result of the experiment is a set of lower and upper limits for the distance between pairs of atoms (constraints). If the number of constraints is sufficient, there will be a finite number of possible conformations of the protein compatible with the data. The more constraints we are able to measure, the more similar to each other will these structures be (Figure 1.17).

The number of constraints in an NMR experiment is strongly dependent on the flexibility of the protein and of its regions in solution: if a given region is very mobile, it will be very difficult to identify the neighbors of its atoms because they will not spend enough time next to each other. In such cases, we cannot measure the interactions but we recover very valuable information about the intrinsic mobility of the protein structure.

Question: How do I evaluate the quality of an NMR structure and how does it compare with X-ray structure?

»NMR structures are usually reported with *rmsd* values (the square root of the average sum of the squared distances between corresponding atoms, see later) for the various structures compatible with the data. The lower the *rmsd*, the

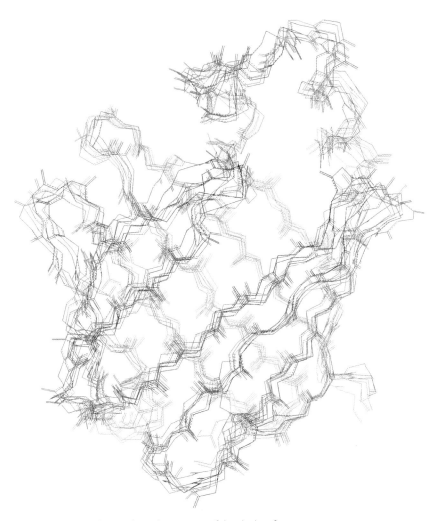

Figure 1.17 Several NMR-derived structures of the chicken fatty acid-binding protein. Note that the exposed regions are less well defined than the central core of the protein.

higher the accuracy of the measurements. The answer to the second part of the question depends on whether the question is posed to a crystallographer or to an NMR spectroscopist! There is no clear and definite answer because the two experiments give different, albeit related, information.«

Question: Does the crystal structure of a protein reflect its "true" native and functional structure?

»This question is often asked. Several lines of evidence point to a positive answer – structures of the same protein solved by both X-ray crystallography and NMR, or solved independently in different crystal forms, are the same within the experimental error. Furthermore, protein crystals are full of solvent (and for this reason very fragile) and it has often been shown that crystallized enzymes can function inside the crystal; they are therefore deemed to have the correct native functional structure.«

1.5
The PDB Protein Structure Data Archive

Structures determined by both X-ray and NMR are deposited in a data base called PDB. X-ray structure entries consist of a single structure; for NMR entries there is a variable number of structures, usually approximately 20, compatible with the data. Each entry is uniquely identified by a four-letter code. In the first part of a PDB entry there are the name of the molecule, the biological source, some bibliographic references, and the R and Rfree factors. There is also information about how chemically realistic the model is, i.e. how well bond lengths and angles agree with expected values (the values found in small molecules). For a good model, average deviations from expected values should be no more than 0.2 Å in bond lengths and 4° in bond angles. The SEQRES records contain the amino acid or nucleic acid sequence of residues in each chain of the macromolecule studied, whereas the HELIX, SHEET and TURN records list the residues where secondary structure elements begin and end and their total length.

Question: Where can I find the sequences of all the proteins of known structure?

»There is also a database of the sequences of known structures, usually called pdb, containing the sequences extracted from the SEQRES records. Be aware, however, that even if some parts of the protein are not visible in the electron density map, because they are too mobile or because the protein was partially degraded in the experiment, their sequence will still be included in the SEQRES field. In other words the sequence

corresponds to that of the studied molecule, not necessarily to
the part of the molecule the structure of which is contained in
the entry. The database of sequences of known structure
called ASTRAL only includes the sequence of the part of the
molecule that has been experimentally determined.«

After this initial part of the file, the actual coordinates are listed in records identified by the keyword ATOM. These include a serial number for the atom, the atom name, the alternative location indicator, used when the electron density for the atom was observed in two positions, the chain identifier, a residue sequence number and code, the x, y, and z orthogonal coordinates for the atom, the occupancy, and the temperature factor. For example the record:

ATOM 1281 N GLY Z 188A 29.353 66.969 17.508 1.00 28.84

describes the nitrogen atom of a glycine unit with residue number 188 and residue code A. The coordinates are $x = 29.353$, $y = 66.969$, $z = 17.508$. The occupancy is 1.00 (i.e. complete) and the B factor is 28.84 (corresponding to an uncertainty in the position of this atom of 0.6 Å.

Question: Which is the minimum occupancy of atoms reported in a PDB file?

»There is no lower limit to the value of the occupancy for an
atom. It can be 0 if the position of an atom was guessed on the
basis of the positions of the surrounding atoms. Be aware that
none of the widely used structure-visualization packages
highlights them automatically. It is always advisable, if one is
working on a particular region of a protein, to verify the B
factor and occupancy of its atoms.«

It is worth briefly describing the residue number and code, because these are often the cause of much frustration when trying to use a PDB file: the residue number is not necessarily consecutive. For example, trypsin is synthesized as a longer molecule the first 15 amino acids of which must be enzymatically removed to produce the active protein. The first residue number in the 3PTI entry for trypsin is indeed 16. A common numbering scheme is occasionally used for a family of evolutionarily related proteins, and in such circumstances the residue numbering follows the scheme. If one of the proteins of the family contains amino acids inserted among the commonly accepted numbering, the residue code is given a letter. In the 3TPI entry, for example, we find:

    ALA 183
    GLY 184A
    TYR 184
    LEU 185
    GLU 186

1 Sequence, Function, and Structure Relationships

GLY 187
GLY 188A
LYS 188

For NMR structures the headers do not, of course, include the R factors and the resolution. The ATOM fields are quite similar, the B factor is usually set to 0 and the sections referring to each of the models are included between the records MODEL and ENDMDL.

1.6
Classification of Protein Structures

Protein structures can be classified according to their similarity, in terms of secondary structure content, fold, and architecture. There are a few widely used

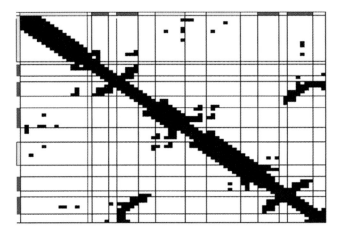

Figure 1.18 A distance map for the domain shown at the bottom (PDB code: 1RUN). The secondary structure of the protein is shown in the first line and first column of the matrix. Black and gray regions correspond to beta strands and alpha helices, respectively. Filled cells correspond to distances shorter than 6 Å.

classifications of protein structures which are extremely useful for navigating through the protein structural space. These are collected and made available to the community via Web servers and differ in the method used to obtain the classifications.

FSSP is a classification method based on comparison of the "distance matrices" of proteins. These are an alternative representation of protein structures (Figure 1.18) obtained by filling a matrix. Each row and each column of a matrix represent an amino acid and each cell contains the distance between the amino acid in the row and the amino acid in the column in the protein structure. Given two proteins, we can compare their distance matrices and derive a structural superposition between their atoms, i.e. the superposition that minimizes the distance between corresponding pairs of atoms. The resulting structural distance between the two proteins, defined as the root mean square of the average sum of the squared distances, is used by FSSP to cluster the known structures and to classify them.

CATH is another classification of protein structures based on use of a different algorithm to compute structural similarity. In this classification the two distance matrices that are compared contain the vectorial distance between pairs of atoms, rather than the scalar one. CATH provides a hierarchical classification of the structures, identifying four levels of similarity – Class, Architecture, Topology, and Homology. The Class is defined on the basis of the predominant type of secondary structure (all alpha, all beta, alpha and beta, and domains with little or no secondary structure). The Architecture describes the overall shape of the domain structure as determined by the orientations of the secondary structures ignoring the connectivity between the secondary structures. It is assigned on the basis of visual inspection of the proteins and of literature data. The Topology level depends on the structural distance between proteins, and evolutionarily related proteins are grouped at the Homology level, on the basis, as we will see, of sequence-based methods.

Finally, SCOP is another classification with a hierarchical organization including Class, Fold, Superfamily, and Family levels. The main Class types in SCOP are all alpha, all beta, alpha plus beta, and alpha/beta. A protein is assigned to one of the classes according its predominant secondary structure. The other classes include multi-domain, membrane, and cell surface proteins and peptides, small proteins, peptides, designed proteins, and low-resolution structures. The second level of classification, Fold, includes proteins with similar topological arrangements for which an evolutionary relationship cannot be identified, the third level (Superfamily) includes proteins that are believed to share a common ancestor. Proteins related by an unambiguous evolutionary relationship are grouped at the Family level. The classification in SCOP is essentially manual, although some automatic pre-processing is used to cluster clearly similar proteins.

None of these classifications is intrinsically better than any other and they usually agree with each other.

1.7
The Protein-folding Problem

The stability of each possible conformation of an amino acid chain depends on the free energy change between its folded and unfolded states:

$$\Delta G = \Delta H - T\Delta S$$

where ΔG, ΔH, and ΔS are the differences between the free energy, enthalpy, and entropy, respectively, of the folded and unfolded conformations. The enthalpy difference is the energy associated with atomic interactions within the protein structure (dispersion forces, electrostatic interactions, van der Waals potentials, and hydrogen bonding that we will describe in more detail later) whereas the entropy term describes hydrophobic interactions. Water tends to form ordered cages around non-polar molecules, for example the hydrophobic side-chains of an unfolded protein. On folding of the polypeptide chain, these groups become buried within the protein structure and shielded from the solvent. The water molecules are more free to move and this leads to an increase in entropy that favors folding of the polypeptide.

Question: What does an unfolded protein looks like?

»Although we generally assume that the unfolded chain is in a random coil conformation, i.e. that the angles of rotation about the bonds are independent of each other and all conformations have comparable free energies, the reader should be aware that, in reality, unfolded proteins tend to be less disordered and more compact than ideal random coils, because some regions of the polypeptide can interact more favorably with each other than with the solvent.«

In a cell proteins are synthesized on ribosomes, large molecular assemblies comprising proteins and ribonucleic acid molecules. Special adaptor molecules, tRNA molecules, recognize a triplet of bases on the messenger RNA, which in turn has been synthesized by following instructions contained in the genome, and adds the appropriate amino acid to the nascent chain. The synthesis of an average protein takes approximately a minute; the time required for folding, i.e. for achieving the "working" native structure, is comparable. Some slow steps of the reaction, for example formation of disulfide bonds, are accelerated by specific enzymes. Other proteins are also involved in the folding process and their role is either to protect the nascent protein chain (shielding the hydrophobic regions that are exposed to solvent before folding occurs) or to provide a more protected environment for folding; there is no evidence that anything but the amino acid sequence determines the native protein structure in vivo.

In the nineteen-sixties the American chemist Christian Anfinsen and his co-workers performed a series of seminal experiments demonstrating that the native

1.7 The Protein-folding Problem

conformation of a protein is adopted spontaneously or, in other words, that the information contained in the protein sequence is sufficient to specify its structure. The enzyme selected by Anfinsen for the experiment was ribonuclease A (RNase A), an extracellular enzyme of 124 residues with four disulfide bonds (Figure 1.19). As already mentioned, these are covalent bonds arising as a result of oxidation of the sulfhydryl (SH) groups of the side-chains of two cysteines, when they are close to each other. The result is an S–S bond between their sulfur atoms. In Anfinsen's experiment, the S–S bonds were first reduced to eight –SH groups (by use of mercaptoethanol, a reducing agent with the chemical formula $HS-CH_2-CH_2-OH$); the protein was then denatured by adding urea in high concentration (8 Molar). (The urea molecule enhances the solubility of nonpolar compounds in water and therefore reduces the strength of the stabilizing hydrophobic interactions that hold the protein structure together.) Under these conditions the enzyme is inactive and becomes a flexible random polymer. In the second phase of the experiment the urea was slowly removed (by dialysis); the –SH groups were then oxidized back to S–S bonds. We expect that if the protein is able to assume its

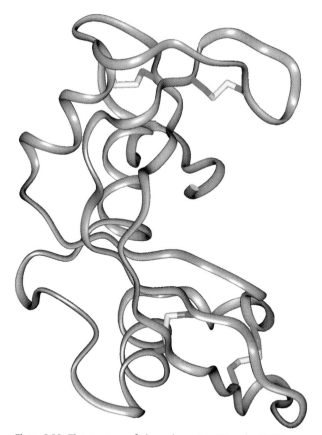

Figure 1.19 The structure of ribonuclease A (PDB code: 1AFK). Note the four disulfide bridges.

correct tertiary structure, the correct pairs of cysteines are close to each other so that the correct disulfide bonds form and the protein regains its activity. Indeed, the refolded protein regained more than 90% of the activity of the untreated enzyme.

> Question: How can we be sure that the protein was really unfolded after adding urea and mercaptoethanol?

»Anfinsen and co-workers also performed a control experiment to demonstrate that RNase A was completely unfolded in 8 Molar urea. RNase A was reduced and denatured as above, but in the second phase the enzyme was first oxidized to form S-S bonds, and only afterwards was the urea removed. If the protein is really in a random conformation in 8 Molar urea, it is likely that the cysteines are in different relative positions in different molecules and will randomly pair giving rise to scrambled sets of disulfide bonds. Because there are eight cysteine residues in ribonuclease, there are $7 \times 5 \times 3 \times 1 = 105$ different ways of forming disulfide bonds, only one of which is correct and leads to the formation of a functional enzyme. The experiment indeed showed that only about 1% of the activity could be recovered in this control experiment. Later the same experiment was successfully repeated using a chemically synthesized protein chain, i.e. a protein that had never seen a cell or a ribosome.«

These experiments demonstrated that proteins can, indeed, adopt their native conformation spontaneously, but immediately raised a fundamental problem known as the Levinthal paradox – if the same native state is achieved by various folding processes both in vivo and in vitro, we must conclude that the native state of a protein is thermodynamically the most stable state under "biological" conditions, i.e. the state in which the interactions between the amino acids of the protein are the most energetically favorable compared with all other possible arrangements the chain can assume. But an amino acid chain has an enormous number of possible conformations (at least 2^{100} for a 100-amino-acid chain, because at least two conformations are possible for each residue). It can be computed that the amino acid chain would need at least $\sim 2^{100}$ ps, or $\sim 10^{10}$ years to sample all possible conformations and find the most stable structure.

Levinthal concluded that a specific folding pathway must exist and that the native fold is simply the end of this pathway rather than the most stable chain fold. In other words, Levinthal concluded that the folding process is under kinetic rather than under thermodynamic control and that the native structure corresponds not to the global free energy minimum but rather to one which is readily accessible. The hypothesis underlying Levinthal's reasoning is that the energetically favorable contacts that stabilize the structure arise only when the chain is folded or nearly folded. In other words, the protein chain must first lose all its entropy (being locked in a given conformation) and, only when the correct conformation is reached, can the entropy loss be compensated by the gain in enthalpy. A wealth of literature addresses the Levinthal paradox and we will not dwell on the details here, except to say that, in general, the paradox can be solved by thinking of the folding process as a sequential process in which the entropy decrease is immediately or nearly immediately compensated by an energy gain and that, in this hypothesis, the time-scale computed for the folding process approximates that observed in nature.

1.8
Inference of Function from Structure

The Structural Genomics Initiatives promise to deliver between 10 000 and 20 000 new protein structures within the next few years and, as we will see in this book, many more protein structures will be modeled. The challenge is obviously to exploit this large amount of structural data to predict the functions of these proteins. Proteins sharing a common evolutionary origin (homologous proteins) have similar structures, as we will see shortly, and, occasionally, proteins that do not seem to share an evolutionary relationship might turn out to share the same topology. One can expect to gain insight into a protein's function from analysis of other, structurally similar, proteins. There are at least three difficulties to be overcome in this process:
- homologous proteins might have originated by gene duplication and subsequent evolution and therefore have acquired a different function;
- some folds are adopted by proteins performing a variety of function; and, finally,
- the protein of interest might have a novel, not yet observed, fold.

What can we learn from the analysis of a protein structure? We can certainly identify which residues are buried in the core of the protein and which are exposed to solvent. The structure will also tell us the quaternary structure of the protein – the structure observed in the crystal is often that which is biologically active, although there are exceptions that might create difficulties.

The presence of local structural motifs with functional roles can be detected by analyzing the structure. For example, the presence of a helix–turn–helix motif suggests that the protein binds DNA. Two alpha helices intertwined for approximately eight turns with leucine residues occurring every seven residues are the dimerization domains of many DNA-binding proteins. A motif in which a zinc atom is bound to two cysteines and two histidines separated by twelve residues is called a zinc finger and is found in DNA and RNA binding proteins. Other shorter and non-contiguous local arrangements can be identified and associated with a function, for example the arrangement of serine, histidine and aspartate in serine proteases (Figure 1.3).

When no known local functional motif can be detected, it is still possible to analyze clefts on the surface of the protein (in more than 70% of proteins the largest cleft contains the catalytic site) and highlight the presence of amino acid side-chains that are likely to be involved in catalytic activity. Biochemical knowledge can help us to postulate a catalytic mechanism.

For non-enzymes, the problem is much harder to solve. Detecting the protein–protein interaction sites is very difficult and there is not yet a completely satisfactory method, although analysis of the hydrophobicity of the surface in conjunction with automatic learning approaches is leading to some success.

When other members of the evolutionary family are known, analysis of the conservation and variability of amino acids facilitate estimation of the functional importance of different parts of the structure. Any approach used to detect function from structure has a major limitation, however: the molecular function of a protein does not tell us very much about its biological role. If we predict a protease activity for an enzyme, even if we can identify the likely substrate, we are still left with the question of its biological role, because these enzymes participate in many processes, from digestion to blood coagulation, from host defense to programmed cell death.

It should also be mentioned that, recently, more and more proteins (called moonlight proteins) have been found to perform more than one function, often totally unrelated to each other. This might be frustrating, but should not be surprising – there is no reason evolution should not take advantage of different surface regions of proteins to endow them with different activities.

Another significant proportion of proteins seem to be intrinsically disordered and assume their native structure only when they meet and bind with their partners (natively unfolded proteins).

> Question: Is the property of being disordered functionally important ?
>
> »The property is often evolutionarily conserved and is, therefore, deemed to be functionally relevant. The reason might be that their flexibility enables these proteins to bind several targets or to provide a large interacting surface in big complexes. This might also be a clever way of engineering high specificity but low affinity. A large interaction surface usually confers both properties but, if the protein has to expend energy for folding before binding, specificity can still be achieved without large affinity. Other explanations can be invoked for this behavior, for example the lifetime of an unfolded protein in a cell is probably shorter and this can provide a regulatory mechanism. More simply, there is no reason why evolution should select against these proteins, because selective pressure acts on the function of the protein and is not concerned with what the protein does when not involved in its functional interactions, assuming it does not have a deleterious effect on the cell.«

The existence of moonlight and natively unfolded proteins makes the problem of inferring the function of novel proteins even more complex and, indeed, this is one of the fields that is attracting more attention at the present. It is easy to predict that many new more powerful methods will be developed in the near future, taking advantage both of the wealth of data that is being accumulated and of novel approaches. The problem is somewhat recent – before the start of structural genomics initiatives, the determination of the structure of a protein was usually the final step of its characterization and was aimed at understanding the details of its functional mechanism or interactions rather than to infer its biological function from scratch. Only recently are we facing the challenge of having an available structure and no functional information.

1.9
The Evolution of Protein Function

In 1859 Darwin published "The Origin of Species", a book that laid the foundation of evolutionary theory. The careful observations he made during his travels led him to realize that the taxonomy of species could be explained by postulating gradual changes occurring generation after generation and to propose that changes might result in competitive advantage for the organism as members of a population better fitted to survive leave more offspring. The traits of successful individuals then become more common, whereas traits that do not increase, or even reduce, the fitness become rarer or disappear altogether. Evolution, therefore, acts to transform species in the direction of better fitness for the environment. Darwin also had the intuition that even traits that do not, by themselves, confer any selective advantage, might become predominant in a population if they attract the preference of sexual partners. At about the same time Mendel discovered that the traits of the partners are not blended in the offspring – on the contrary, specific characters are sorted and inherited. The foundations for a molecular theory of evolution only needed identification of the material carrying the characteristics, i.e. the DNA, and this happened approximately half a century later.

At the time it did seem surprising that a simple molecule such as DNA, which is, after all, only a polymer comprising a limited set of different nitrogen-containing bases (only four, as it happens) each attached to a sugar and a phosphate group, could explain the diversity between an amoeba and a man. Only fifty years later, however, the diffraction data collected by Rosalind Franklin enabled James Watson and Francis Crick to build a structural model of DNA. The structure of this molecule immediately suggested how DNA can be replicated and copied. This is probably the only example in history in which knowledge of the structure of a macromolecule has immediately provided information about a novel functional mechanism.

What remained to be understood was how the DNA could code for proteins, i.e. what was the code linking the four-character alphabet of a DNA molecule with the twenty-letter alphabet of proteins. The path that led to unraveling this code was much harder than the previous steps, it took years of study and experiments to obtain the genetic code (Table 1.2), i.e. the correspondence between each triplet of bases of the DNA and the coded amino acid. With rare exceptions, the genetic code is universal, it is used by bacteria, plants, animals, more proof, if needed, of the theory of evolution.

Table 1.2 The genetic code.

		U	C	A	G	
U		Phe	Ser	Tyr	Cys	U
		Phe	Ser	Tyr	Cys	C
		Leu	Ser	STOP	STOP	A
		Leu	Ser	STOP	Trp	G
C		Leu	Pro	His	Arg	U
		Leu	Pro	His	Arg	C
		Leu	Pro	Gln	Arg	A
		Leu	Pro	Gln	Arg	G
A		Ile	Thr	Asn	Ser	U
		Ile	Thr	Asn	Ser	C
		Ile	Thr	Lys	Arg	A
		Met	Thr	Lys	Arg	G
G		Val	Ala	Asp	Gly	U
		Val	Ala	Asp	Gly	C
		Val	Ala	Glu	Gly	A
		Val	Ala	Glu	Gly	G

A story related to the genetic code is very instructive for highlighting that biology is a complex field and that beautiful elegant theories might fall short of describing how evolution has shaped life. Francis Crick was one of the contributors to the long and labor-consuming path that led to the discovery of the genetic code. In 1957, however he devised the following solution to the problem of mapping the four DNA bases to the twenty amino acids. We have four bases and twenty amino acids. There are only 16 possible combinations of two characters out of an alphabet of four (4^2), therefore a coding system that associates a pair of bases with each amino acid could only encode sixteen amino acids. If we use a triplet of bases to code for an amino acid, then we have 64 (4^3) possible combinations. The problem is how to map 64 triplets to twenty amino acids. Let us assume that only a subset of the 64 possible triplets codes for amino acids and that they are such that, when two are placed next to each other, only one "reading frame" can be meaningful. For example, if the codons CGU and AAG code for an amino acid, then none of the triplets GUA and UAA can be coding sequences, so if the DNA contains the sequence CGUAAG there is no ambiguity in how it should be read and translated.

In this hypothesis, none of the triplets AAA, CCC, GGG, and UUU can be coding, because if they were the sequences AAAAAA, CCCCCC, GGGGGG and UUUUUU would be ambiguous, so we are left with 60 codons. Next, only one of the codons that are cyclic permutations of each other can be coding. Let us consider the codons ACG, CGA and GCA. Ambiguities arise if more than one of these is used. For example, if we use ACG and CGA, the sequence ACGACG is ambiguous. This implies that we can only select one codon every three; we are therefore left with only 20 out of the 60 codons. Crick and colleagues did realize that, apart for its elegance and simplicity, there was no other support for this hypothesis and indeed they wrote: "We put it forward because it gives the magic number – 20 – in a neat manner and from reasonable physical postulates." The theory was very elegant and equally wrong, demonstrating that evolution selects a working alternative, and does not seem to be interested in elegant minimal design!

Each cell of an organism, with rare exceptions, carries a complete copy of the genetic material. Bacteria usually have only one copy of the genetic material organized as a circle made of double-stranded DNA. More complex species have cells with nuclei and reproduce sexually. Most of the DNA in such organisms resides in the cell nucleus and is arranged in several chromosomes that occur in pairs. One member of each pair comes from the mother and one comes from the father. Some DNA is also stored in separate organelles within the cell, called mitochondria and chloroplasts.

Darwin viewed evolution in terms of the genealogical relationships among species or major groups of organisms over a long time span. The impressive progress in molecular biology enables us to study evolution in molecular terms, by looking at the change in the genetic make-up of a population and at the differences between species in terms of difference in their DNA sequence.

Replication, either in the process of creating new somatic cells (mitosis) or in the process of creating germ cells (meiosis), is extremely accurate, and there are several mechanisms to ensure its fidelity; errors are inevitable, however. Environmental factors, for example high-energy radiation, can, moreover, cause random damage to the DNA molecule. Mechanisms exist for repairing the damage, but sometimes they introduce errors. These can be of two types – replacements of DNA bases by others or deletions or insertions of any number of bases. A base replacement may or may not affect a protein sequence. The change may occur in an intron or in another region of the DNA that does not code for a protein. When it occurs in a protein–coding region, the replacement might lead to a codon that is translated into the same amino acid as the original, because of the redundancy of the genetic code. Alternatively, an amino acid residue in the original protein may be replaced by a different amino acid in the mutated protein (missense mutation) or the mutation can involve a stop codon. If a codon for an amino acid residue is changed to a stop codon, the protein will be terminated prematurely and will usually be non-functional (nonsense mutation) whereas if a stop codon mutates into a codon for an amino acid residue the translation continues, elongating the amino acid chain until the next stop codon is encountered.

Large insertions or deletions in the coding regions of a protein almost always prevent production of a useful protein. Short deletions or insertions in a coding region of any number of bases other than a multiple of three usually have a drastic effect – they cause a shift in the reading frame during translation, resulting in a meaningless change in the amino acid sequence in the C-terminal direction from the point of mutation. When the insertion or deletion involves multiples of three bases, it does not affect the sequence of the protein outside the site of the insertion or deletion and may or may not affect its function.

A gene, or a whole chromosomal region, might be duplicated, leading to a situation in which two copies of the same gene are present. If there is no selective pressure, the two copies may evolve independently – one copy may continue to code for the protein performing the original function whereas the other may evolve by mutation into an entirely different protein with a new function. New combinations of existing genes are occasionally produced at the beginning of meiosis when the chromatids, or arms, of homologous chromosomes break and reattach to different chromosomes (crossing-over). It is easy to see how these mechanisms can account for variability within a species and differences between different species.

One can also speculate about the mechanisms by which new species arise. A species is defined as a set of individuals that, in the wild, would mate and produce fertile offspring. A new species can therefore originate when some individuals, for whatever reason, do not mate with the rest of the population for a sufficient length of time. These individuals can follow a different evolutionary path that might result in genetic incompatibility with the original group. This can be because of physical separation between groups of individuals, or acquisition of different lifestyle, or spreading of the individuals over a huge geographical range. Study of evolution and of the relationships between species and their proteins is of paramount importance in modern biology; most of what we can infer about the function and structure of biological elements comes from analysis of their differences and similarities with the corresponding elements in different species. The possibility of comparing the sequence of entire genomes has also resulted in the possibility of highlighting which parts of the genome are under evolutionary pressure and are, therefore, deemed to be functionally important. It is not surprising that a plethora of tools and theories has been developed to highlight evolutionary relationships, some of which will be described in this book in the appropriate context. Here we will review some elements of the terminology that are commonly used.

Phylogeny is an inferred pattern of evolutionary relationships between different groups of organisms. Usually we depict a phylogeny as a rooted tree in which the length of the branches is proportional to the divergence time and each leaf represents a species. We also use a tree representation to indicate the evolutionary relationships between genes or proteins. In this representation each leaf is a gene or a protein and the lengths of the branches are proportional to the accumulated changes between the molecules. It should be kept in mind that, although molecular trees derived from protein sequences are related to phylogenetic trees, the former refer to the observed difference, for example in functional regions, and do not, therefore, necessarily relate to a proper phylogenetic tree, because different evolutionary pressure can result in different rates of evolution for different genes or proteins. For example, the rate of mutation of hemoglobin is approximately one change per site every billion years whereas fibrinopeptides can accumulate nine time this number of mutations in the same period of time.

1.9 The Evolution of Protein Function

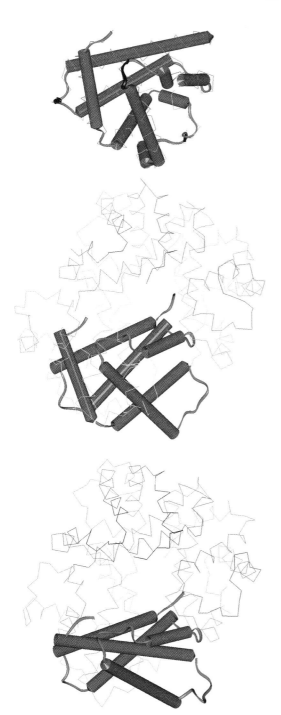

Figure 1.20 Three homologous chains: myoglobin, and alpha and beta globin. Note that these three structures are similar and evolutionarily related, but they are paralogous, i.e. they have arisen by gene duplication.

Sometimes molecular trees are unrooted, i.e. they depict differences, but make no hypothesis about the location or properties of the ancestor elements. For some applications, such a tree is good enough, but – once again – it is not a tool to derive evolutionary times.

Two elements (whether genes, proteins, anatomical structures) that derive from a common evolutionary ancestor are called "homologous". In anatomy and in protein analysis homology does not guarantee common functionality – the anterior wings of a bat and a human arm are homologous but have a different function. If two anatomical parts resemble each other, they are called analogous. The usual example here is the eye of vertebrates and that of the squid, they seem similar and have a similar function, but have a different evolutionary origin. We will call two protein structures analogous if they resemble each other (i.e. they have the same topology) but there is no evidence of common evolutionary origin. When we refer to genes or proteins, the concept of homology must be further specified. Two proteins (or genes) believed to have diverged from each other because of speciation events are called orthologous whereas two proteins (or genes) that are homologous, i.e. derived from a common ancestor, but have arisen after a duplication event are called paralogous. This is by no means a semantic distinction but one of the major issues in protein bioinformatics, because it has a very relevant impact on the prediction of the biological function of newly discovered proteins.

Let us use an example to illustrate the issue. Myoglobin is a monomeric protein the function of which is to store oxygen in the muscle. Human and chimp myoglobin are orthologous – they descend from the same ancestral protein via speciation and they also have the same function. Hemoglobin, the tetrameric oxygen transporter is composed by two pairs of alpha and beta chains. Myoglobin, alpha hemoglobin, and beta hemoglobin are all homologous, they descend from a common ancestor, as is apparent from their structural similarity (Figure 1.20), but they are paralogous, because they have arisen by the duplication of an ancestral globin gene. If two homologous genes are found in the same genome, it is easy to see that they are paralogous. Paralogous genes can, however, also be present in different genomes – human myoglobin and chimp alpha globin are paralogous. As illustrated by this example, because of the possibility of paralogous relationships, the finding that two genes are homologous does not necessarily imply they share the same function.

1.10
The Evolution of Protein Structure

If a base-substitution event occurs in a protein-coding region of a genome, the net effect can be the substitution of one residue of the encoded protein with a different one. What is the effect on the structure of the protein of a single amino acid replacement? There are only two possibilities. One is that the fine balance between the gain and loss of free energy of folding is compromised, there is no single global energy minimum for the new sequence, and it does not fold any more. Because proper folding is required for function, the most likely outcome is that the organism is not viable and the mutation is not propagated in the population. The second alternative is that the energy landscape of the new sequence changes, but it still contains a free energy global minimum and the corresponding native structure is still able to perform the same function as the original protein.

How likely is it that the new conformation is very different from the original? Statistics and physics both tell us this is extremely unlikely and that the most probable outcome is that the new sequence assumes a structure very similar to that of the original protein. In other words the substituted amino acid is accommodated into the structure with only local perturbation and without dramatic global changes in structure and function. Indeed substantial changes in protein architecture, because of a single, evolutionarily accepted mutation, have not yet been observed. Therefore, when residue substitutions and short insertions/deletions accumulate in members of an evolutionarily related family of proteins, they will cause local

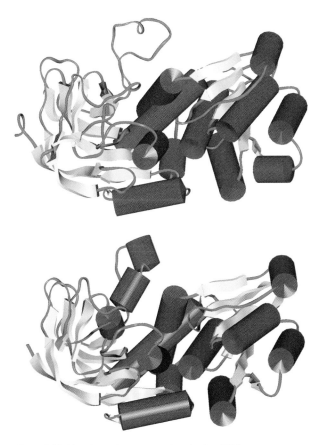

Figure 1.21 Two homologous proteins sharing 30% sequence identity: alcohol dehydrogenase from horse liver (PDB code: 1YE3) and *Acinetobacter calcoaceticus* (1F8F).

structural perturbations without affecting the general shape, or topology, of the protein. Of course, the greater the number of mutations (or, equivalently, the further the proteins are in the evolutionary scale), the larger will be the difference between the protein structures. We can quantify this qualitative observation by measuring the relationship between the different sequences of homologous proteins and their structural divergence, as we will see in the next section. We will merely mention here that it is accepted that pairs of evolutionarily related proteins sharing at least 30% sequence identity have a similar fold (Figure 1.21).

> *Question: Is the sequence-to-structure relationship limited to naturally evolved proteins or does it reflect an intrinsic property of amino acid sequences?*
>
> »It is important to understand that the sequence–structure relationship already described occurs when there is the requirement, as in natural evolution, that each step of the evolutionary path must be functional. If we were to artificially introduce several mutations into a protein structure without guaranteeing that each step yields a functional mutant, we might be able to change its structure dramatically. Indeed, in 1994 two scientists (Creamer and Rose) issued the "Paracelsus Challenge" – they offered a prize of $1 000 to the first group to successfully convert one protein fold into another while retaining at least 50% sequence identity with the original fold (this challenge was named after Paracelsus, the 16th century Swiss alchemist). At least three groups have published reports in response to this challenge.«

Insertions and deletions, even when they do not cause frame shifts, are difficult to accommodate in a protein structure because a substantial part of the molecule, the internal core, is tightly packed. The periodicity of secondary structure elements also implies that insertion or deletion of one amino acid can change the pattern of interaction of the whole region very markedly. Not surprisingly, most of the observed amino acid insertions and deletions are located at the surface of the protein structure and outside secondary structure elements.

Fusion of two or more initially independent genes leads to the production of multidomain proteins with new combinations of functions in a single protein. In eukaryotes, this process is thought to be facilitated by the presence of introns (intervening sequences in genes that are not coding). These represent regions in which genes can easily be recombined – if one exon from a gene coding for a protein region with a given function is inserted into an intron region of a gene for a protein carrying a different function, the new hybrid protein might be capable of both functions and serve a new physiological role. It should be mentioned, however, that there is no evidence that exons preferentially encode structural or functional units.

1.11
Relationship Between Evolution of Sequence and Evolution of Structure

To analyze the relationship between sequence and structural divergence in quantitative terms, we must define a measure of distances in sequence and structure space. Several unsolved problems connected with this issue will be discussed later in this book. For the moment let us assume we know how to find the correspondence between the amino acids of two evolutionarily related proteins that reflects their evolutionary history. In other words, let us assume that, given two proteins, we can construct a matrix such as that shown in Figure 1.22 in which the first row contains the amino acid sequence of the first protein, possibly with inserted spaces, and the second contains the amino acid sequence of the second protein, again possibly including spaces. Two amino acids in the same column are assumed to originate from the same amino acid of the ancestor protein. The spaces represent insertion and deletion events. Given this correspondence, called alignment, we can define the distance in sequence space simply as the fraction of amino acids that is different between the two proteins.

Sequence 1	...	M	Q	D	G	T	S	R	F	T	C	R	G	K	–	–	P	I	H	H	F
Sequence 2	...	L	S	E	G	N	H	K	L	S	C	R	H	D	Q	G	P	V	N	D	Y

Figure 1.22 Alignment of two fragments of the sequences of the proteins shown in Figure 1.21.

To measure distance in structure space, we use the root mean square deviation (*rmsd*) between corresponding atom pairs after optimum superposition. In practice, we apply the rigid-body translation $T = (T_x, T_y, T_z)$ and rotation $R = (R_x, R_y, R_z)$ to one of the proteins that minimizes the value:

$$rmsd(T, R) = \min_{T,R} \sqrt{\frac{1}{N} \sum_{i=1}^{N} \left[(x_i - R_x x'_i + T_x)^2 + (y_i - R_y y'_i + T_y)^2 + (z_i - R_z z'_i + T_z)^2 \right]}$$

The set of corresponding atom pairs should, once again, be that which reflects the evolutionary correspondence between amino acids. Obviously, when we are superimposing two proteins with different sequences, we can only use atoms that are common between any two protein structures, for example the Cα, or the atoms of the backbone, or the backbone plus the Cβ for all amino acids except glycine.

As already discussed, peripheral parts of the proteins can undergo local rearrangements that can be quite substantial. If there are insertions and deletions it is obviously impossible to compute the *rmsd* values for the inserted residues, but even if this is not so, the fact that the superposition procedure minimizes the sum of the squares of the deviations means the *rmsd* is dominated by the most divergent regions and, if we include them, the changes in the more conserved regions would be masked. This leads to the need to superimpose separately the conserved "cores"

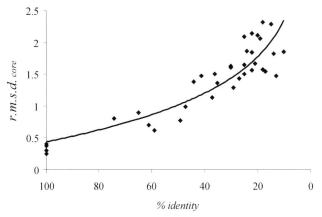

Figure 1.23 Relationship between sequence identity and structural similarity. The plot is obtained using the same set of proteins originally analyzed by Lesk and Chothia.

of evolutionarily related proteins and, therefore, calls for a definition of the core. Many empirical procedures are commonly used. Chothia and Lesk, for example, propose the following: given two related proteins, one first superimposes the main chain atoms of corresponding elements of the secondary structure and then continues to add residues at either ends of the elements until the distance between the alpha carbons of the last added residue deviates by more than 3 Å. Next, one jointly superimposes these "well fitting" regions and calculates the resulting *rmsd*.

Now that we know how to measure sequence distance and structural divergence, we can investigate the relationship between them. In a seminal paper Chothia and

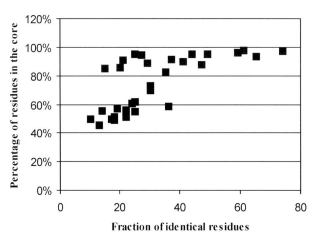

Figure 1.24 Relationship between sequence identity and the extent of the common structural core between pairs of homologous proteins. (Data from the original Chothia and Lesk analysis on thirty-two pairs of proteins).

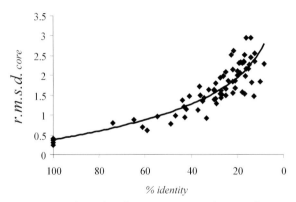

Figure 1.25 Relationships between sequence identity and structural similarity. The plot was obtained by using a larger set of proteins than in Figure 1.23, but the trend is essentially the same.

Lesk selected thirty-two pairs of homologous proteins of known structure, identified the common core within each pair with the procedure described above, and computed the *rmsd* values between the core of each pair as a function of the sequence identity between the two protein sequences (Figure 1.23). Their conclusions were that, as sequences diverge, the extent of the common core between two homologous proteins decreases. The common core contains almost all the residues when pairs of closely related proteins (with sequence identity >50%) are considered; when residue identity drops below 20% the structures might diverge quite substantially and the core can contain as little as 40% of the structure (although in some cases it can include most of the structure) (Figure 1.24). They found that the relationship between the structural divergence of the core and the sequence identity was in accordance with the equation:

$$rmsd_{core} = 0.40 e^{\frac{(100 - \%identity)}{100}}$$

Although the original analysis by Chothia and Lesk was limited to 32 pairs of proteins only, a relationship with very similar parameters was obtained when the analysis was repeated using a much larger sample (Figure 1.25).

Suggested Reading

These two books guide the reader through a fascinating tour of protein architecture and show how the shape of a protein is linked to its function:

A.M. Lesk (**2001**) Introduction to Protein Architecture, Oxford University Press

A.M. Lesk: (**2004**) Introduction to Protein Science: Architecture, Function, and Genomics, Oxford University Press

Another excellent book that describes the principles of protein structure, with examples of key proteins in their biological context, is:

C.-I. Branden, J. Tooze (**1999**) Introduction to Protein Structure, 2nd edn, Garland, New York

Readers interested in learning more about crystallography and nuclear magnetic resonance spectroscopy can consult these two seminal books:

J. Drenth (**1994**) Principles of Protein X-ray Crystallography, Springer

K. Wüthrich (**1986**) NMR of Proteins and Nucleic Acids, John Wiley and Sons

Every biochemistry book contains at least a chapter dedicated to the structure and function of proteins. Readers can read the relevant chapters from:

J.M. Berg, L. Stryer, J. L. Tymoczko (**2002**) Biochemistry, 5th edn, W. H. Freeman

D.L. Nelson, M. M. Cox (**2004**) Lehninger Principles of Biochemistry, 4th edn, W. H. Freeman

D. Voet, J. Voet (**2004**) Biochemistry, 3rd edn, Wiley

The original paper describing the PDB data archive (http://www.pdb.org) is:

F.C. Bernstein, T. F. Koetzle, G. J. B. Williams, E. F. Meyer Jr, M. D. Brice, J. R. Rodgers, O. Kennard, T. Shimanouchi, M. Tasumi (**1977**) The Protein Data Bank: A Computer-based Archival File for Macromolecular Structures. J. Mol. Biol. **112**, 535–542

Structural classification of proteins, SCOP (http://scop.mrc-lmb.can.ac.uk/scop), is described in:

A.G. Murzin, S. E. Brenner, T. Hubbard, C. Chothia (**1995**) SCOP: a structural classification of proteins database for the investigation of sequences and structures. J. Mol. Biol. **247**, 536–540

CATH (http://www.biochem.ucl.ac.uk/bsm/cath/) C. A. Orengo, A. D. Michie, S. Jones, D. T. Jones, M. B. Swindells, J. M. Thornton (**1997**) CATH – A hierarchic classification of protein domain structures. Structure **5**, 1093–1108

FSSP (http://www.bioinfo.biocenter.helsinki.fi:8080/dali/index.html) L. Holm, C. Sander (**1996**) Mapping the protein universe. Science **273**, 595–602

The original paper by Anfinsen is:

C.B. Anfinsen, E. Haber, M. Sela, F. White Jr (**1961**) The kinetics of formation of native ribonuclease during oxidation of the reduced polypeptide chain. PNAS **47**, 1309–1314

It is difficult to gain access to the original paper in which Levinthal described the paradox that took his name:

C. Levinthal (**1969**). In: P. Debrunner, J. C. M. Tsibris, E. Munck (Ed.) Mossbauer spectroscopy in biological systems, University of Illinois, Urbana, IL, pp. 22–24

However, several more recent papers describe the reasoning explained in the paper.

The worldwide initiatives in structure genomics are described at http://www.rcsb.org/pdb/strucgen.html where a good collection of informative background information about this project is also available.

I would also recommend reading "The Origin of Species" by Charles Darwin, because of its outstanding historical and scientific interest.

The paper by Francis Crick and colleagues, describing their theory on the comma free genetic code is:

F.H.C. Crick, J. S. Griffith, L. E. Orgel (**1957**) Codes without commas. PNAS USA **43**, 416–421

Finally, the analysis of the relationship between sequence and structural similarity in proteins by C. Chothia and A. M. Lesk can be found in:

C. Chothia, A. M. Lesk (**1986**) The relation between the divergence of sequence and structure in proteins. EMBO J. **5**, 823–826

2
Reliability of Methods for Prediction of Protein Structure

2.1
Introduction

Imagine a scenario in which a biologist has obtained the amino acid sequence of a previously unknown protein and is trying to gather as much information as possible on the molecule. The first thing he or she will do is to access the internet and try to find and use the many tools available, including those, which we will discuss later, that produce three-dimensional models of the protein of interest. Undoubtedly, if more than one method is used, the results will not coincide. Occasionally, the discrepancies between the models obtained will be small, but for more difficult structures even the same method will present the user with more than one solution to the problem. The issue is, therefore, how do we select the best model, i.e. the one we will use as a theoretical framework to investigate further our biological system?

Given more than one model, we will certainly try to see which of the proposed structures is more consistent with available experimental information. Before investing our time in this enterprise, however, we would like to know which of the results are expected to be more reliable, by looking at the quality of the results that each method has produced in the past. Furthermore, as we will discuss later, it is possible that none of the available methods can produce a model sufficiently good for our purposes.

This chapter is devoted to description of how the accuracy of a method is evaluated and what should we look at when using it.

Two methods are usually used to estimate of the quality of a method. In the first we select a set of cases for which the answers, for example the three-dimensional structures, are known, pretend we do not know them and verify how similar the results of the method are to the real experimental answer. In the second, we use the method to predict the structure of a protein that is not yet known, but that will soon be elucidated, and, when the data are available, compare the predicted and observed features. Both strategies have advantages and disadvantages, which we will discuss; before this, however, we must address the problem of how we can measure the difference between the predicted and experimental structures.

Prediction of a protein structure can be limited to prediction of the location of secondary structure elements, to production of an alignment to a protein of known structure, or to a fully fledged three-dimensional model of the protein.

Protein Structure Prediction. Edited by Anna Tramontano
Copyright © 2006 WILEY-VCH Verlag GmbH & Co. KGaA, Weinheim
ISBN: 3-527-31167-X

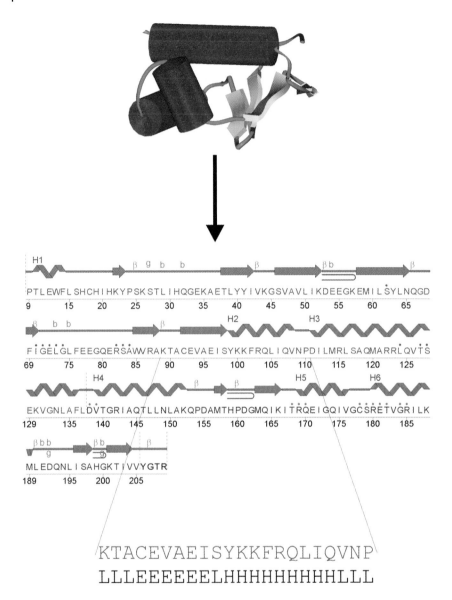

Figure 2.1 The secondary structure, predicted or experimental, of a protein can be encoded as a string. Usually only alpha helices (H) and beta strands (E) are considered. 3(10) helices are included in the helical definition, and every other conformation is indicated with L (loop). The figure shows the three-dimensional structure of an SH3 domain, a very useful summary of its properties that can be found at the site http://www.ebi.ac.uk/thornton-srv/databases/pdbsum/, a database providing an overview of every macromolecular structure deposited in the Protein Data Bank, and the string encoding the secondary structure of a fragment of the protein.

2.2
Prediction of Secondary Structure

We can envisage the results of a secondary structure prediction method as a string of characters representing, for example, residues predicted to be in alpha helices, beta strands, and irregular regions. This string must be compared with the experimentally determined secondary structure of the protein which can also be encoded as a string (Figure 2.1).

It is, however, a fact that elements of regular structure in proteins are not as regular as we would like them to be, especially at their termini. For example, an alpha helix is defined both by its φ and ψ backbone angles and by its typical hydrogen-bond pattern. It is quite common that the last residue of the helix does not form all its hydrogen bonds or even that, if the helix has a bend, some internal residues lose one or two hydrogen bonds (Figure 2.2). Similarly, in beta sheets

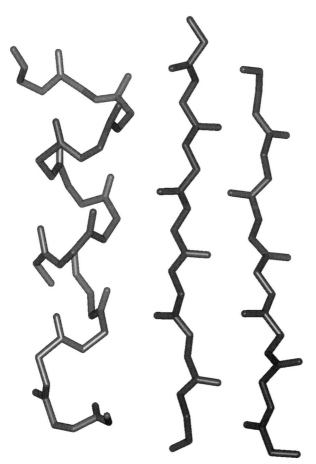

Figure 2.2 Helices and strands contain irregularities highlighted by the absence of the standard hydrogen-bond pattern.

there might be bulges (i.e. a region of irregularity in a beta sheet, in which the normal pattern of hydrogen bonding is disrupted by insertion of one or more extra residues) or the strands might move slightly further away from each other at their ends and lose the regular pattern of hydrogen bonds that characterizes them (Figure 2.2).

Several methods are widely used to assign the limits of the secondary elements; the two most commonly applied are probably DSSP and STRIDE. The first is essentially based on the pattern of main chain hydrogen-bonds between amino acids; the second uses empirical energetic criteria to define the presence and/or absence of hydrogen-bonds and also takes into account the values of the φ and ψ angles. Other methods are based on Cα–Cα distances or on the curvature of the peptide backbone. Each of these methods has its drawbacks. That, in general, these methods do not all agree on secondary structure assignment for a given protein is simply because there is no right answer.

Whereas a few years ago this was a negligible problem in secondary structure prediction, because the accuracy of the methods was lower than the aforementioned experimental uncertainty, as methods become increasingly clever and accuracy reaches very respectable values, we need to keep in mind that small differences between methods might not be significant because they are within experimental uncertainty.

Given the two strings representing predicted and observed secondary structure, we usually use two terms to evaluate their agreement – the Q index and the SOV score. The Q index is simply defined as:

$Q = 100 \times$ (Number of residues correctly predicted in a given state)/(Total number of residues)

In secondary structure prediction most methods assign only three possible states – helix, strand, and other and the Q index (called $Q3$) is the sum of the Q indexes for each of the three states, as shown by the example in Figure 2.3.

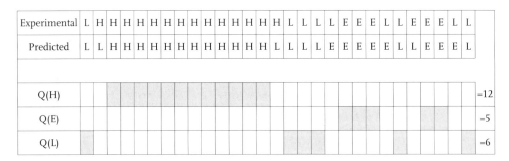

$Q3 = 12/29 + 5/29 + 6/29 = 0.79$

Figure 2.3 The Q3 value is computed by adding the percentage of residues correctly predicted to be in the helical, strand, or loop conformation.

2.2 Prediction of Secondary Structure

The *SOV* measures the overlap between segments, rather than the accuracy of the prediction of the state of a single residue. For one of the states, for example the helices, we define:

$$SOV(i) = \sum_i \frac{Minov(S1_i, S2_i) + \Delta(S1_i, S2_i)}{Maxov(S1_i, S2_i)} \times len(S1_i)$$

Minov ($S1_i$, $S2_i$) is the length of the regions in which two corresponding secondary structure segments overlap (two elements are defined as corresponding if they have at least one residue in common). *Maxov* is the length of the region in which either has residues in the secondary structure and:

$$\delta(S1_i, S2_i) = min\left((Maxov(S1_i, S2_i) - Minov(S1_i, S2_i), Minov(S1_i, S2_i), int\left(len\tfrac{S1_i}{2}\right), int\left(len\tfrac{S2_i}{2}\right)\right)$$

The *SOV* for the three states is computed as the sum of the three *SOV* measures divided by the sum of the lengths of the experimental segments (Figure 2.4).

	L1			H1												L2			E1			L3	E2			L4		
Experimental	L	H	H	H	H	H	H	H	H	H	H	H	H	H	L	L	L	L	E	E	E	L	L	E	E	E	L	L
Predicted	L	L	H	H	H	H	H	H	H	H	H	H	H	H	L	L	L	L	E	E	E	E	L	L	E	E	E	L
Minov L1																												=1
Maxov L1																												=2
Minov H1																												=12
Maxov H1																												=14
Minov L2																												=3
Maxov L2																												=5
Minov E1																												=3
Maxov E1																												=5
Minov L3																												=1
Maxov L3																												=3
Minov E2																												=2
Maxov E2																												=4
Minov L4																												=1
Maxov L4																												=2

Experimental:
len(L1) = 1; len(H1) = 14, len(L2) = 4; len(E1) = 3; len(L3) = 2; len(E2) = 3; len(L4) = 2
Predicted:
len(L1) = 2; len(H1) = 12, len(L2) = 4; len(E1) = 5; len(L3) = 2; len(E2) = 3; len(L4) = 1

δ(L1) = int(min(2−1,1,int(2/2),int(1/2))) = int(min(1,1,1,0.5)) = 0
SOV(L1) = ((1+0)/2)*1 = 0.5

δ(H1) = int(min(14−12,12,int(12/2),int(14/2))) = int(min(2,12,6,7)) = 2
SOV(H1) = ((12+2)/14)*14 = 14

δ(L2) = int(min(5−3,3,int(4/2),int(4/2))) = int(min(2,3,2,2)) = 2
SOV(L2) = ((3+2)/5)*4 = 4

δ(E1) = int(min(5−3,3,int(5/2),int(3/2))) = int(min(2,3,2.5,1.5)) = 1
SOV(E1) = ((3+1)/5)*3 = 2.4

δ(L3) = int(min(3−1,1,int(2/2),int(2/2))) = int(min(2,1,1,1)) = 1
SOV(L3) = ((1+1)/3)*2 = 1.34

δ(E2) = int(min(4−2,2,int(3/2),int(3/2))) = int(min(2,2,1.5,1.5)) = 1
SOV(E2) = ((2+1)/4)*3 = 2.25

δ(L4) = int(min(2−1,1,int(2/2),int(1/2))) = int(min(1,1,1,0.5)) = 0
SOV(L4) = ((1+0)/2)*2 = 1

SOV = (0.5+14+4+2.4+1.34+2.25+1)/29 = 0.88

Figure 2.4 The Figure shows how to calculate the SOV value for a prediction. One SOV value is computed for each segment of secondary structure. In the example there is one helical segment, two strands, and three loops. For the helix, *maxov* is 14, *minov* is 12 and the lengths of the predicted and observed helical segment are 14 and 12, respectively. Thus delta, the integer minimum between ((*maxov-minov*), *minov*, 14/2, 12/2) is 2. The resulting value for SOV(H1) is 14. For the strands, we must compute an SOV value for each segment. The first has *maxov* = 5, *minov* = 3, and the lengths of the experimental and predicted regions are 3 and 5, respectively. This implies that the value of delta for this segment is 1 and its SOV is 2.4. For the second strand, *maxov* = 4, *minov* = 2, delta = int(min(2,2,1,1) = 1 and therefore SOV = 2.25. At the end, the SOV values for each segment are summed and divided by the length of the sequence.

2.3
Prediction of Tertiary Structure

Comparison of the structures of two different proteins requires solution of both the alignment and superposition problems – we must find which pairs of atoms we want to superimpose and then find the translation and rotation of one of the two molecules that minimizes the *rmsd* between the sets of paired atoms. When comparing a model and its experimental structure the alignment problem is trivial, because the two structures have the same sequence, but this does not mean there are no other problems. Consider the example shown in Figure 2.5. On the left is the superposition between a structure and its respective model obtained using all the Cα atoms; on the right the superposition is computed using about a half of the Cα atoms.

The inspection of the figure might convince the reader that the superposition shown on the right is more representative of the overall quality of the model, because it shows how well the packed core of the structure is predicted, and does not take into account the deviations in peripheral parts of the structure. The superposition shown on the right can also be very informative for the end user if, as in this example, it involves the set of atoms that are part of the active site of the molecule. The fraction of superimposed structure and the *rmsd* values are, obviously, correlated and this implies that, given two different predictions for the same structure, we need to compare the *rmsd* values for an equivalent fraction of superimposed atoms and, at least in principle, the *rmsd* values for the biologically important regions should be considered more relevant. A useful way of obtaining a qualitative impression of the relative accuracy of two models is to plot the *rmsd* values as a function of the fraction of superimposed atoms. In the example shown

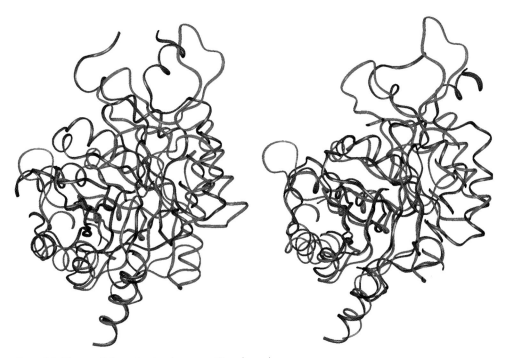

Figure 2.5 The result from structural superposition depends both on the number of superimposed atoms and the *rmsd* value.

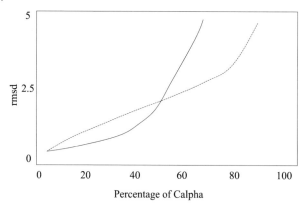

Figure 2.6 Given a model and an experimental structure, the *rmsd* values for the best superposition can be computed by superimposing 5%, 10%, 15%, etc. of the model and target structures. Two models can be compared by plotting the *rmsd* as a function of the fraction of superimposed atoms. In the example, the model represented by the continuous line fits better than the model represented by the dotted line, when approximately one half of the structure is considered. The situation is inverted when larger fragments are superimposed.

in Figure 2.6, it is clear that the model indicated by the continuous line is closer to the experimental structure for about one half of the structure whereas the other has a better overall quality.

Ideally we would like to know whether the well modeled part of the first model includes the regions we are most interested in, for example the active site or a binding region. Unfortunately there is no way of easily obtaining this information automatically and analysis of the model by the end user is required. Another issue we must take into account is that the *rmsd* is a quadratic measure and therefore it weights large deviations between the structure and the experimental structure relatively more than smaller deviations. This poses a problem: if a loop is incorrectly predicted in two different models, do we really want to penalize more the model that placed it farther away from the right answer or do we just want to regard both answers as incorrect if the loop deviates from the correct answer by more than a given acceptable threshold? The latter is probably more reasonable and, indeed, it has become increasingly common not to use *rmsd* as a measure of distance between the backbone atoms of a model and a structure, but rather to set a distance threshold and count how many corresponding atoms are within the threshold. To be more general, we may count the number of atoms within several distance thresholds and average them. A frequently used measure is the GDT-TS value defined as:

¼ ((Fraction of Calpha atoms within a 1 Å distance) + (Fraction of Calpha atoms within a 2 Å distance) + (Fraction of Calpha atoms within a 4 Å distance) + (Fraction of Calpha atoms within an 8 Å distance))

Of course, different distance thresholds can be used.

The GDT-TS measure gives an estimate of the correctness of the overall structure but does not consider, for example, how well the side chains are modeled. The measure that is more often used for assessing the quality of side-chain predictions is based on comparison of the dihedral angles of the side chain, usually limiting the calculation to, at most, the first two angles, called χ_1 and χ_2 (Figure 2.7) and computing the fraction for which the deviation is lower than a given threshold, usually 30°.

The next issue we must consider is the quality and possible uncertainties in the experimental structure. In different cases, one might wish to exclude some subsets of atoms from the superposition, for example:

- Atoms with high B factor. If the model is reasonably close to the experimental structure, it is advisable not to consider atoms the B factors of which are very high, because this indicates their position is not well defined.
- Solvent-exposed side chains. Exposed side chains, especially if they are long, are usually mobile.
- Regions involved in crystal contacts. Regions involved in crystal contact between different molecules in the crystal might have a conformation different from that in solution.
- Side-chains with ambiguous chi values. Occasionally rotation of 180° around some side-chain bonds might be undetectable by X-ray crystallography, for example the rotation around the Cα alpha and Cβ beta bond of valine (affecting the computed value of the χ_1 angle) around the Cβ Cγ gamma bond of histidine and asparagine (affecting χ_2), or around the Cγ–Cδ bond of asparagine (affecting χ_3).

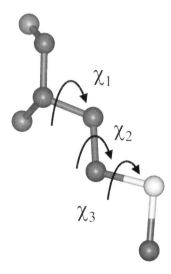

Figure 2.7 Definition of the side-chain χ angles. The amino acid shown here is methionine.

2.4
Benchmarking a Prediction Method

Benchmarking is the process of comparing methods, services, and products in a specifically defined and strictly controlled environment. In structure prediction, this translates into selecting a set of cases with known answers and using them to evaluate the accuracy of a prediction method. Most structure prediction methods are based on heuristics, they use data that must be estimated using a set of proteins for which the structure is known (training set). The accuracy of the predictions on the training set is expected to be higher than on a different test set. Because homologous proteins have similar structure, the quality of the predictions will also be higher on proteins homologous with proteins in the training set. Therefore the test set should contain proteins that share no significant sequence similarity with the training set and, if the test is to be balanced, both training and test sets should have a similar distribution of structure classes and types.

Other biases might be more subtle and difficult to avoid. The length, composition, and cellular localization of a protein might have an effect on its structure. Many methods use information derived from the evolutionary history of the protein under examination, so the number of known homologous proteins in the evolutionary family might be another piece of information that should be taken into consideration when selecting the appropriate test and training sets.

Because the number of proteins of known structure is limited, rather than using separate training and test sets, structure prediction methods often employ cross-validation or jack-knife techniques. In a jack-knife test of N proteins, one protein is removed from the set, the parameters are estimated from the remaining $N-1$ proteins, and then the structure of the removed protein is predicted and the accuracy of the prediction is measured. This process is repeated N times by removing each protein in turn. If the estimation step is very time-consuming, a more limited cross-validation can be performed by splitting the sample into M subsets, where $M < N$.

Figure 2.8 Benchmarking consists in selecting a subset of the data to derive parameters and use the remaining data to test the method. The data are split into a test set (dark gray) and a training set (light gray). The training set is used to derive the results and these are tested on the test set. Next, a different non overlapping set of test data is selected, and their properties are predicted using parameters derived from the remaining data. The procedure is repeated until all data have been used once as test sets. The accuracy of the method is the average accuracy on the different test sets.

Parameters are divided from $(M-1)N$ proteins and tested on the remaining N proteins. This process is repeated M times, once for each subset (Figure 2.8).

2.5
Blind Automatic Assessments

Benchmarking is not always possible, especially if the training procedure is very time consuming. It is, furthermore, useful to evaluate different methods on the same test set to be able to compare their results. For methods that can be run automatically, the solution is to set up an automatic system that collects the results of different methods as soon as a new protein structure is determined, and therefore before any method had a chance to use it in the training set. This has the added advantage that the methods can use all available data in their training set and are not restricted to deriving parameters from a more limited training set.

EVA is one server that performs this useful service to the community. Every day, EVA downloads the newest protein structures from the PDB, extracts the sequences for every protein chain, and sends them to each prediction server registered for the experiment. The results collected are then evaluated and made public. EVA covers several methods that predict solvent accessibility, secondary structure, and complete three-dimensional modeling. The proteins used in the experiment are such that no pair of them has more than 33% identical residues over more than 100 residues aligned.

Livebench is another continuous benchmarking server, but it limits itself to the evaluation of three-dimensional models of proteins not sharing a significant sequence similarity (and therefore deemed to be non-homologous) to any protein of known structure. Every week, new entries in the PDB database are submitted to participating servers. Every week the results are collected and evaluated. In this experiment a target is skipped if it is shorter than 100 residues or longer than 500 residues. The evaluation uses only the Cα positions of the models. It first performs a rigid-body superposition of the model and the structure and then computes the maximum number of atoms within 3 Å after superposition.

The results of both servers are publicly available via the Internet and are extremely useful tools that should be consulted before using any prediction server.

2.6
The CASP Experiments

EVA and Livebench can be used to evaluate the performances of automatic servers, but these are not the complete scenario of prediction methods. First, there is at least the hope that human intervention, with exploitation of data from the target protein, can improve the models. Second, there are methods that cannot easily be automated, either because they are in early stages of their development, or because they require manual evaluation of the results.

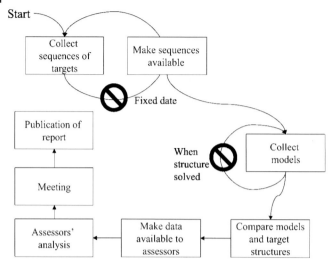

Figure 2.9 The CASP experiment runs every two years. In the spring, approximately, targets are collected from experimenters working on the resolution of their structure. The sequences are made available to predictors who can submit predictions until the structure is solved. Numerical comparison of models and targets is performed by a group of scientists led by John Moult and Krzystof Fidelis. The data are then passed to thee assessors, chosen by the community on the basis of their expertise, who analyze the data and try to derive general conclusions about the state of the art in the prediction field. In approximately December of the same year, predictors, assessors, and organizers convene in a meeting to discuss the results and, later, publish the final reports in the scientific journal *Proteins: Structure, Function and Bioinformatics*.

In 1994 John Moult proposed a world-wide experiment named CASP (critical assessment of techniques for protein structure prediction) aimed at establishing the current state of the art in protein structure prediction, identifying what progress has been made, and highlighting where future effort may be most productively focused. The organization of the experiment is depicted in Figure 2.9.

Crystallographers and NMR spectroscopists who are about to solve a protein structure are asked to make the sequence of the protein available, with a tentative date for release of the final coordinates. Predictors produce and deposit models for these proteins (the CASP targets) before the structures are made available. CASP also tests publicly available servers on the same set of targets providing a unique opportunity to verify the effectiveness of human intervention in the modeling procedure. A panel of three assessors compares the models with the structures as soon as they are available and tries to evaluate the quality of the models and to draw some conclusions about the state of the art of the different methods. The task is divided among assessors in such a way that one looks at models of proteins that share significant sequence similarity with a protein of known structure, one looks at those sharing a significant structural similarity, but no clearly detectable sequence similarity with proteins of known structure, and the third evaluates all the remaining models. It is expected that the first set of targets is predicted using a technique called comparative modeling and the second using fold-recognition methods; because the experiment is run blind, however, i.e. the assessors do not know who the predictors are until the very end of the experiment, it is entirely possible that different techniques are used by different groups for the same target. We will use this division in the rest of this book to describe the various approaches to structure prediction, although, as we will discuss, there is substantial overlap between them.

The results of the comparison between the models and the target structures are discussed in a meeting where assessors and predictors convene, the conclusions are made available to the whole scientific community via the World Wide Web and by publication of a special issue of the journal *Proteins: Structure, Function, and Bioinformatics*. New categories, namely prediction of function, domain boundaries, and disordered regions have recently been included, but we will not discuss their results here. The CASP experiment has been extremely successful; it has been repeated every two years since it was first inaugurated and there is no sign it is going to be discontinued in the near future (Figure 2.10).

It has several merits. First of all, it has raised the issue of objective evaluation of structure prediction methods prompting the development of the continuous automatic assessment methods already described and fostering the development of similar initiatives in other fields, for example prediction of protein–protein interaction, gene finding, and scientific literature mining. It has also been instrumental in the development of common formats and evaluation measures in the field. It does also have limitations, however. Predictors in CASP are not necessarily in an ideal position to produce the best models, because of the time limitation imposed by the experiment. Also, because the results are public and very visible, predictors might not try "risky" innovations.

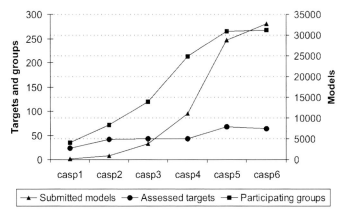

Figure 2.10 The plot shows the numbers of targets, participating groups, and models submitted to each of the editions of CASP from 1994 (CASP1) to 2004 (CASP6). All the thousands of models are publicly available on the CASP web site.

The assessors change every time, with rare exceptions, and are free to analyze the data in different ways and draw their own conclusions, although a set of measures is provided by the CASP organizers and is available for all the targets of all the editions. The most important are the GDT-TS values already described, *rmsd* values for several subsets of superimposed atoms, and the number of correctly aligned residues between target and model (a residue is considered correctly aligned if, after superposition of the experimental and modeled structure, its Cα atom falls within 3.8Å of the corresponding experimental atom, and there is no other Cα atom of the experimental structure that is nearer).

We will often refer to the CASP results in this book. We will also discuss how the CASP results can be used to measure progress between different editions, a non trivial issue, because it requires comparison of results obtained on a different set of targets, and at different times, therefore taking advantage of data bases of sequences and structures of very different size.

Suggested Reading

Description of most of the evaluation data are available on the CASP Web site (http://predictioncenter.org) where all the results of comparisons between models and target structures can be found.

The original definition of *SOV* has been published in:

B. Rost, C. Sander, R. Schneider (**1994**) Redefining the goals of protein secondary structure prediction. J Mol Biol **235**, 13–26

A modification of the definition has been proposed in:

A. Zemla, C. Venclovas, K. Fidelis, B. Rost (**1999**) A modified definition of Sov, a segment-based measure for protein secondary structure prediction assessment. Proteins, **34**, 220–223

The web sites of EVA and Livebench are:
http://cubic.bioc.columbia.edu/eva/
http://bioinfo.pl/LiveBench/

The special issues of the journal "Proteins: Structure, Function and Bioinformatics" describing the results of the CASP experiment are published every two years by Wiley–Liss.

3
Ab-initio Methods for Prediction of Protein Structures

3.1
The Energy of a Protein Configuration

The native protein structure is the lowest free energy conformation that can be achieved kinetically by the polypeptide. It therefore seems natural to face the problem of predicting the structure of a protein by computing the free energy of every possible conformation and selecting the structure corresponding to the global free energy. The problem can be divided into two sub-problems – evaluation of the free energy of a given conformation and the search strategy for finding all possible conformations. Because the number of possible conformations of a protein is, as already mentioned, enormous, exhaustive search strategies cannot be employed; one must, instead, perform an approximate search strategy, the better the sampling of lower energy conformations, the better the search method.

In the next few pages we must discuss and understand some basic aspects of protein physics and, believe it or not, it is easier to do so by looking at equations. It is not always necessary to understand all the details of an equation – for example capturing the meaning of the Schrödinger equation requires a profound understanding of quantum-mechanics, and this is certainly not required to predict a protein structure. Hopefully, however, the reader will be able, by looking at the equations, to understand which variables are involved and the difficulty of the computation involved, without being distracted, or troubled, by the details of the formulas.

3.2
Interactions and Energies

When protein conformational energy is discussed, people talk almost indifferently about forces and energies. Although this can be confusing, remember that the two entities are directly related – the force is the derivative of the energy.

The behavior of a molecule can be completely described by the Schrödinger equation, the quantum mechanical equivalent of Newton's laws and of the law of conservation of energy in classical mechanics. The equation states that:

Protein Structure Prediction. Edited by Anna Tramontano
Copyright © 2006 WILEY-VCH Verlag GmbH & Co. KGaA, Weinheim
ISBN: 3-527-31167-X

$$i\hbar \frac{\partial \Psi}{\partial t} = -\frac{\hbar^2}{2m}\frac{\partial \Psi}{\partial x^2} + V(x)\Psi(x,t) \qquad (1)$$

where i is the imaginary unit, $\hbar$ is a constant, m is the mass of the particle, V is the potential, and Ψ the wave function. A wave function is a scalar function that describes the properties of waves – the Schrödinger equation predicts the future behavior of a dynamic system in terms of the probability of future events. What is important to understand is that to describe a system in terms of its energy we must consider all nuclei and electrons and their interactions. Even if we take into account the great difference between the masses of the electrons and nuclei, and treat them separately (Born-Oppenheimer approximation), it is impossible to solve the Schrödinger equation for systems larger than a few atoms. In practice, the equation is of relevance to protein structure prediction only insofar it can be solved for reasonably small systems and results from these model systems provide data for more approximate energetic calculations.

If we could solve the Schrödinger equation for reasonably large systems such as a protein and a set of surrounding water molecules, we could at least be certain that the energy of each conformation is computed correctly and we would be left "only" with the problem of exploring the conformational space of the system. In practice this is impossible. We need to find a function that approximately describes the energy of the interactions that occur in a protein molecule using a simplified representation of both the system (with the atoms being represented by points located at the center of their nuclei) and of the energetic contributions of each interaction in the protein. These interactions can, by and large, be divided into two types – covalent and non-bonded.

3.3
Covalent Interactions

A covalent bond between two atoms is formed if they share electrons (one pair in single bonds and two pairs in double bonds), but the effect is not localized and the electron density increase has an effect on the rest of the molecule. The standard way of approximating the potential energy for a bond is to treat the bond as a spring between the two atoms and describe its energy by use of Hooke's law (Figure 3.1):

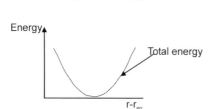

Figure 3.1 The classical approximation for a chemical bond.

$$E_{bond} = K_r(r-r_{eq})^2 \qquad (2)$$

where r is the length of the bond, r_{eq} the equilibrium bond length, and K_r the spring constant (which is higher the stronger the bond).

Use of this equation is justified by the observation that bonds between chemically similar atoms in a wide variety of molecules have similar lengths (e.g. bonds from carbon atoms to hydrogen atoms are approximately 1 Å) so we can assume that the observed equilibrium value corresponds to the minimum potential energy.

Any function around its minimum value can be expanded as a Taylor power series:

$$f(x) = f(x_0) + \frac{\partial f(x_0)}{\partial x}(x-x_0) + \frac{1}{2}\frac{\partial^2 f(x_0)}{\partial x^2}(x-x_0)^2 + \frac{1}{6}\frac{\partial^3 f(x_0)}{\partial x^3}(x-x_0)^3 + \ldots \qquad (3)$$

If we deform a bond around its equilibrium value, i.e. its energy minimum, the first derivative will be zero. Near the equilibrium, the expansion will be dominated by the term in x^2, therefore a function that varies around its minimum can be approximated by a quadratic function, $f(x) \approx A + Bx^2$. Because we are not interested in absolute values, the constant A can be neglected and this brings about the Hooke's law.

To approximate the energy needed to stretch a bond around its equilibrium value, we need to know the equilibrium length and the spring constant. The first is usually derived by analysis of small molecule X-ray crystal structures, the second by optimization or from quantum calculations on model systems. More accurate approximations can occasionally be used, but Hooke's law is usually used in most energy-evaluation procedures in protein structure prediction. It should be clear that this approximation can be used to compute the energy difference between length distances around their equilibrium value and has nothing to do with the energy of formation or breaking of a bond, in which case a quantum mechanical treatment is required.

Similar approaches can be used approximate the energy difference of variation of bond angles (Figure 3.2):

$$E_{bond} = K_\theta(\theta-\theta_{eq})^2 \qquad (4)$$

For dihedral angles matters are slightly more complex, because they do not have a single energy minimum – indeed, they are usually represented by (Figure 3.3):

$$E_{dihedral} = \sum_{n=1}^{N} K_\phi[1 + \cos(n\phi-\gamma)] \qquad (5)$$

where N is the number of minima and γ is the angular offset.

In practice, it is found that this potential is not sufficient to represent the energy of a dihedral angle and often a non-bonded energy interaction term between the first and last atoms of the quadruplet is combined with Eq. (5).

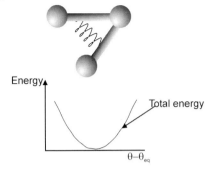

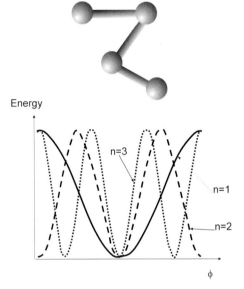

Figure 3.2 The classical approximation for the angle between three bonded atoms.

Figure 3.3 The approximation for a dihedral angle.

3.4
Electrostatic Interactions

The molecular scale is dominated by electromagnetic interactions. A nucleus and its electrons interact according to Coulomb's law (Figure 3.4):

$$E = \frac{q_i q_j}{4\pi\varepsilon_0 \varepsilon_r r_{ij}} \tag{6}$$

where q_i and q_j are the charges, r_{ij} their distance, ε_0 the permittivity, and ε_r the dielectric constant of the medium.

As we said, we cannot solve the Schrödinger equation to find the positions of nuclei and electrons, so we usually assign a "formal" charge to an atom, without explicitly considering its nucleus and its electrons. Some amino acids at physio-

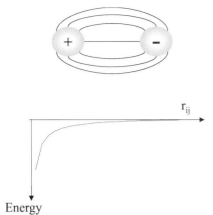

Figure 3.4 The Coulomb interaction.

logical pH are neutral, but some are not and, furthermore, many proteins bind metal ions and the Coulombic interactions must be taken into account. As already mentioned, pairs of residues with opposite charges can form salt bridges that are strong interactions. These interactions are quite rare, especially in the core of proteins, because of the need for electrostatic interaction to compensate the solvation energy, i.e. the energy required to transfer a charge from a polar to a nonpolar solvent. Groups which carry no formal electrical charge can still be polarized, i.e. the orbitals can be distributed in such a way that parts of the molecule carry a charge. Some atoms have a tendency to attract electrons, and are, therefore, electronegative, whereas others have a tendency to lose electrons. The classical example of this is the water molecule, in which the electronegative oxygen attracts the electron and leaves the hydrogen atoms with net positive charges. Two water molecules can therefore form a strong electrostatic interaction, the hydrogen-bond, a bond approximately 2.8Å long and weaker than a covalent bond. Hydrogen bonds, as already discussed, are extremely important in protein structure. The tendency of main-chain carbonyl oxygen bonds to form hydrogen-bonds with main-chain amino groups leads to the possibility of forming different secondary structures; many side-chain groups can also form hydrogen-bonds.

Computationally, treatment of electrostatic interactions requires that an appropriate charge is placed at the position of the nucleus. These charges are partial to take into account the various effects neglected when using a classical approximation and their interaction is computed using the Coulomb's law.

> *Question: How do we compute the partial charge of an atom?*
>
> »The specific charges for each atom are computed by performing quantum mechanical calculations on model systems.
>
> This approach is clearly a very crude approximation, and, indeed, calculation of electrostatic interactions is one of the

weakest points in the classical treatment of macromolecular interactions. Obviously the electric field generated by a charged atom polarizes other atoms and this, in turn, affects the charge of the atom.«

A further complication in the treatment of electrostatic interactions arises because proteins are not in a vacuum. As already mentioned, Coulomb's law contains the term ε_r, the dielectric constant of the medium. The dielectric constant is a macroscopic entity derived from the average microscopic effect of polarization. If we place two opposite charges in a polar medium, the molecules of the medium will tend to line up with the electric field with their dipole such that the positive part points towards the negative charge and vice versa. Their dipole will oppose the electric field, effectively reducing its strength, and thereby reducing the interaction energy between the two charged atoms (Figure 3.5). The dielectric constant takes this effect into account, in the hypothesis that the space around the charges is uniformly filled with a large number of molecules of the medium, but when we deal with proteins the distances between charges is of the same order of magnitude

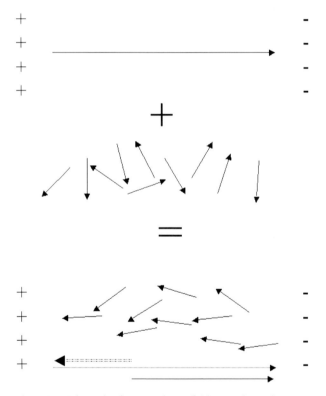

Figure 3.5 Polar molecules in an electric field orient themselves. The net effect is reduction of the electric field that is taken into account, macroscopically, by the dielectric constant.

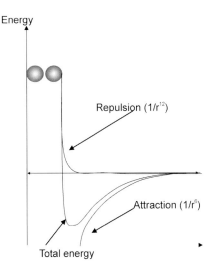

Figure 3.6 The van der Waals radius originates from interplay between the attractive and repulsive forces between two atoms.

as the size of the microscopic dipoles. The uniform distribution hypothesis breaks down and it is no longer possible to simply scale down the interaction by a fixed factor that takes into account the polarizability of the medium.

Sometimes, especially in the past, when computing power was limited, energy calculations were performed using a dielectric constant varying with the distance between the two charges (to take into account the fact that the farther away are two charges the more dipoles are between them), but this is a very crude approximation. Nowadays, it is more common to explicitly include in the calculation a large number of molecules of the solvent and use a dielectric constant of 1 or 2 (to take into account the electronic polarizability).

Electromagnetic interactions also affect uncharged atoms – they vibrate and this produces a dipole moment (because at any given instant the nucleus will be off center relative to the cloud of electrons) that interacts with the similarly generated dipoles of surrounding atoms. This produces an attracting interaction that has been shown to be in accordance with Eq. (7) (Figure 3.6):

$$E_{\text{dispersion}} = \frac{-B_{ij}}{r_{ij}^6} \qquad (7)$$

where B_{ij} depends on the pair of atoms involved and is usually estimated empirically from data derived from small-molecule X-ray crystal structures.

The other effect that must be taken into account is that the orbitals of the atoms cannot overlap because of the Pauli exclusion principle that states that two electrons cannot have the same quantum state. The consequence is that two atoms cannot come too close to each other. This effect can be approximated by assuming that each atom is a hard sphere with a specific radius (the van der Waals radius) and two atoms cannot come closer than the sum of their radii. More realistically, although still approximately, this energy term is modeled as (Figure 3.6):

$$E_{\text{repulsion}} = \frac{A_{ij}}{r_{ij}^{12}} \qquad (8)$$

where A_{ij} is an empirically derived term.

3.5
Potential-energy Functions

If we now take into account all we said about the interactions in protein molecules, we can write the potential energy of a given conformation C as:

$$E_C = \sum_{\text{bonds}} K_b(b_C-b_{eq})^2 + \sum_{\text{angles}} K_\theta(\theta_C-\theta_{eq})^2 + \sum_{\text{dihedrals}} \frac{K_\phi}{2}[1+\cos(n\phi_C-\gamma)] + \sum_{\substack{\text{nonbonded}\\ \text{atoms}}} \left[\frac{A_{ij}}{r_{ij,C}^{12}} - \frac{B_{ij}}{r_{ij,C}^{6}} + \frac{q_i q_j}{\varepsilon_0 r_{ij,C}} \right] \qquad (9)$$

This function gives the approximate value of the potential energy of a protein conformation. It does not include kinetics contributions to the energy and, to compute it, we must estimate a large number of terms such as the equilibrium values, the spring constants, etc.

> Question: What Is the effect of using this approximate equation for the energy of a protein?
>
> »The equation can be used to compute the approximate difference between two different conformations but gives no information about the free energy of forming that conformation and cannot be used to simulate quantum mechanical events such as chemical reactions, bond formation or breaking, etc.«

3.6
Statistical-mechanics Potentials

The atom-based approach to evaluation of protein potential energy is approximate and it has been shown on many occasions, including in the CASP experiments, that its energy terms are not sufficiently accurate for distinguishing the native conformation of a protein from an ensemble of reasonable alternatives. It is useful for quickly exploring some regions around the native conformation of a protein, or to evaluate whether a conformation is physically reasonable, but it still requires a fair amount of computation. An even more approximate, but nevertheless rather useful, approach is a residue-based strategy in which all atomic interactions between residues are attributed to a single point within each residue. This approach is knowledge-based, i.e. the energy contribution of each interaction is estimated from statistical analysis of the known structures of proteins. Calculation of the potentials is based on the heuristic observation that there is a correlation between the frequency of observing a particular structural feature, for example an interaction between two amino acids, and its energy:

$$\text{Frequency (feature)} \propto e^{-\beta E(\text{feature})} \quad (10)$$

This implies that we can count how many times we observe a given feature in the collection of proteins of known structure and, from this, compute its energy. At first sight, this seems to be nothing other than an application of the Boltzmann distribution:

$$p_i = \frac{e^{-\frac{E_k}{KT}}}{\sum_i e^{-\frac{E_k}{KT}}} \quad (11)$$

which states that, in a system at equilibrium, the probability of observing a state i is related to its energy E_i. In the equation, K is the Boltzmann constant and T the absolute temperature. Boltzmann's law is only applicable to an ensemble of identical but distinguishable particles in thermodynamic equilibrium. A database of protein structures consists of many different particles and it certainly cannot be regarded as a system at equilibrium. One way to try and understand why the equation is valid anyway is to think that if an interaction between two amino acids is energetically favorable there will be many protein sequences that can form that interaction. If we observe an interaction very frequently (i.e. in many different proteins with different sequences and structures), we can assume that the interaction is energetically favorable.

Knowledge-based potentials are derived by statistical analysis of known protein structures by counting how many times an interaction occurs. We can, for example, count how many times we observe a pair, for example, alanine–valine, at a distance of 3, 4, and 5 Å from each other. In practice we must introduce another term, their distance along the sequence. This is because we cannot expect our two amino acids to be at a distance incompatible with the number of peptide bonds connecting them. Our potential will have the form:

$$E(a,b,d) = -KT \ln p(a,b,d)/p_0(a,b,d) \quad (12)$$

where p and p_0 are the observed fractions of total contacts observed between the two amino acid types a and b (for example alanine and valine), at a distance d (for example 5 Å) from each other in the known protein structures and in a reference state, respectively. In other words, $p_0(a,b,d)$ represents the fraction of contacts that we expect to observe by chance alone. We need to count the occurrences of all possible pairs of residues, for different through-space and through-bond separations, and normalize these values in respect of a random ensemble of amino acids

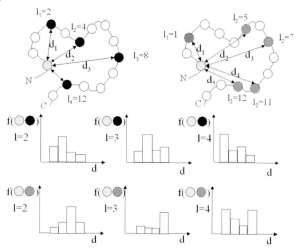

Figure 3.7 Schematic illustration of the method for deriving pair potentials. The number of contacts between the amino acid indicated by the black circle and that represented by the gray circle is counted for each of their separation distances in the sequences (the *l* values in the figure). This is repeated for each structure in the database and for each pair of residues and the frequencies are tabulated as a function of the type of amino acid, the separation distance, and the through-space distance. These values must be normalized by dividing them by the corresponding values observed for the background distribution of random structural arrangements.

(Figure 3.7). As often occurs in computational biology, defining the expected random distribution is the most difficult aspect, because we have no experimental information about the unstructured conformation of our protein.

We can generate many conformations with the same sequence as our target protein and count the observed interactions in this random ensemble, or we can reshuffle our sequence in the same target structures several times and count how many times the observed interactions arise by chance.

Question: How much does this choice affect the result?

»This choice is not without effect. If we use a set of random conformations, they will not, in general, show any secondary structure. Real proteins have a reasonable fraction of amino acids in secondary structures and the interactions corresponding to their geometry will show up more frequently than expected. For example, in alpha helices, residues separated by four bonds are at a fixed distance so that, if we compare a real protein structure distribution with that of a random polypeptide, we will observe a frequency peak at the corresponding distance for residues separated by three intervening amino acids. If, instead, we use real protein structures with reshuffled sequences as our reference distribution, they will also contain secondary structure. The helical pair-wise interactions in our protein structure will be more frequent than expected only for those amino acids that are in a helix more often than for a random sequence.«

When our potential have been derived, we can use it to evaluate the energy of a given conformation by adding the contributions of the observed interactions.

3.7
Energy Minimization

We have discussed how we can evaluate, at least approximately, the energy of a given protein conformation. We will discuss later in this chapter the limitations of our approximations. First, however, we will review the methods that are at our disposal to explore the conformational space of a protein, i.e. to modify the starting structure to obtain a structure with lower computed energy. Let us imagine we can explore every possible conformation of a protein and compute its energy using a potential energy function or a knowledge-based potential. A graph of the energy as a function of the conformation represents the energy landscape of the protein (Figure 3.8) and, if our computed energy were exact, the native structure would correspond to the conformation for which the energy has a global minimum:

E(native conformation) $\leq E$(conformation) for all conformations

The problem of finding the global minimum of the energy function is difficult and, in general, no method can guarantee to find it, but there are techniques able to search for the local minima, i.e. for minima within a certain range from the

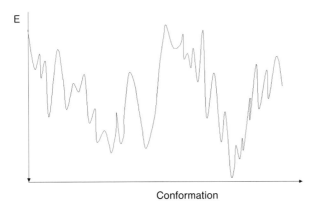

Figure 3.8 Each conformation of a protein has an associated energy, The graph shows a hypothetical plot of the energy as a function of protein conformation. In reality, each conformation is defined by a large number of coordinates, not just one as suggested by the figure.

starting value of the data (the initial conformation in our case). If we have a model of our protein we believe to be rather close in structure to the native conformation, energy minimization procedures could be used to vary the coordinates of the atoms within a certain range and find if any explored new arrangement has an energy lower than that of the starting structure. The approximations introduced in our treatment of the energy are too crude to guarantee that the computed energy difference between two similar conformations reflects their true energy difference with sufficient accuracy, therefore energy minimization is usually only used to remove unfavorable contacts or regularize hydrogen-bond lengths in a model, but there is no guarantee that our energy-minimized structure is closer to the native structure than our starting conformation.

3.8
Molecular Dynamics

Each atom of a protein has a potential energy and therefore feels a force exerted on it equal to the spatial derivative of this potential energy. We can compute the force acting on each atom and, within our classical approximation of a protein, simulate its motion by integrating the Newton's second law of motion:

$$F_i = m_i \frac{\partial^2 x}{\partial t^2} \tag{13}$$

We can calculate the positions of each atom along a series of extremely small time steps and the resulting series of snapshots of structures over time is called a trajectory. What we do in practice is to select a temperature T and compute an initial atom velocity distribution that conforms with the total kinetic energy of the system according to the equation:

$$f = \left(\frac{m}{2\pi KT}\right)^{2/3} e^{-\frac{mv^2}{2KT}} \tag{14}$$

which gives the fraction f of particles of mass m and velocity v in a system with its temperature T. To integrate the Newton equation, we calculate the acceleration, the velocity, and the new positions of each atom at each time step:

$$a_i(t) = \frac{F_i}{m_i}; v_i(t + \Delta t) = v_i(t) + a_i(t)\Delta t; r_i(t + \Delta t) = r_i(t) + v_i(t + \Delta t)\Delta t \tag{15}$$

The time step must be sufficiently small to guarantee that the acceleration (i.e. the force) is practically constant during the step. An adequate time step for a molecular dynamics simulation is of the order of one femtosecond (10^{-15} s). The approximate time scales of the different motions in proteins are listed in Table 3.1. A simulation of a protein should be long enough to sample the motion several times and therefore the length of our simulation must be directly related to the time scale of the phenomenon we wish to simulate.

Table 3.1 Time scale of different types of motion in a protein structure.

Motion	Femtoseconds (or number of steps in a typical molecular dynamics simulation)	Time (s)
Bond stretching	10	10^{-14}
Angle bending	100	10^{-13}
Rotating CH_3 groups	1 000	10^{-12}
Water tumbling	20 000	2×10^{-11}
Chemical reaction	1 000 000 000	10^{-6}

Molecular dynamics simulations at relatively high temperature can be used to explore larger fractions of the protein conformational space while avoiding energetically "unreasonable" conformations, and they are, indeed, often used with this purpose by collecting snapshots of the simulation at regular time intervals and minimizing their energy.

3.9
Other Search Methods: Monte Carlo and Genetic Algorithms

Molecular dynamics samples the conformational space of a protein by moving along a trajectory, although this is not the only option. We can move from one conformation to another by random sampling and accept or reject the new conformation according to some energy-based rule. For example (Monte Carlo Metropolis algorithm), we can start from a conformation with potential energy E and make a random move, for example change a bond angle. We now compute the energy of the new conformation E'. If E' is lower than E, we accept the move, otherwise we extract a random number R between 0 and 1. If R is less than $\exp((E' - E)/KT)$ we keep the new conformation, otherwise we reject it. T is called temperature by analogy with the original applications of the method, but is just a control term and it does not necessarily have physical meaning.

In Monte Carlo methods we accept any change in conformation that reduces the energy, but do not necessarily reject the others. The probability of accepting a move that increases the potential energy depends on how much higher the potential energy of the new conformation is compared with that of the previous conformation, and on the selected temperature of the system. In this way we can move through the conformation space of the protein also exploring regions where the energy is higher, but not "too much higher", than that of our starting conformation (Figure 3.9).

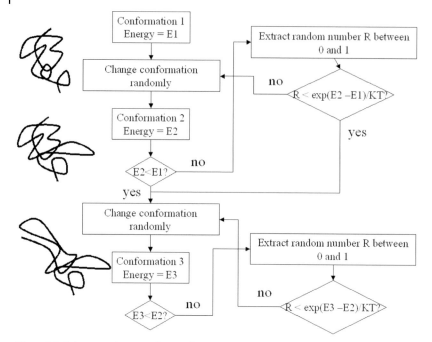

Figure 3.9 Schematic diagram of a simple Monte Carlo simulation.

Physically, our strategy is meant to ensure that the distribution of states at the beginning and at the end of our simulation have the same Boltzmann distribution, i.e. that the probability $p(i)$ of observing the conformation i of energy $E(i)$ is given by the equation:

$$p(i) \propto e^{\frac{-E(i)}{KT}} \tag{16}$$

The "temperature" of the simulation can be reduced after a number of steps, "cooling" the system. This process, called simulated annealing, can be thought of as an adiabatic approach to the lowest energy state.

Another search strategy is based on genetic algorithms that mirrors what nature does during evolution. In this algorithm linear strings of letters (representing the genome) are allowed to mutate, crossover, and reproduce. A genetic algorithm requires the definition of the initial population and of a fitness function. The initial population is usually encoded as a string of bits and the simulation starts by randomly applying "operations" to each individual. For example, a mutation involves exchanging the value of a bit (from 0 to 1 or from 1 to 0), a variation means incrementing or decrementing its bits by a small value (for example the string 0111 represents the number 7, by subtracting 1 from its value we obtain 6 which is encoded as 0110), a crossover means exchanging parts of one individual with parts of another. The next step involves ranking the individuals according to their fitness and, for example, selecting the N with the highest fitness for a new round of simulation. The process is repeated until the desired distribution of fitness is reached, or after a predetermined number of steps. After several generations the population will consist of individuals that are well adapted in terms of the selected fitness function (Figure 3.10). The method does not guarantee that the final population contains the optimum solution for the given fitness function, but the procedure is more efficient than a random search.

Genetic algorithms are often used in protein structure prediction to generate a small set of native like conformations. The initial population is formed by several conformations of a protein and the fitness function can be the potential or knowledge-based energy, but we need a formalism to represent protein structures. We can make the genetic algorithm work on numbers rather than bits and encode our protein using the coordinates of the atoms, but in this case the probability that a mutation originates a chemically impossible protein structure is very high. More common is to encode the protein as a list of numbers representing the ϕ and ψ dihedral angles of its backbone, leaving bond distances and Ω angles fixed. In this representation the mutation operator requires replacement of one torsion angle with another (usually selected from among the values most frequently observed for the given residue), the variation operator will increment or decrement the angle by a preselected value, for example 5 or 10°, whereas crossover requires exchanging an N-terminal or a C-terminal, or an internal fragment between two conformations.

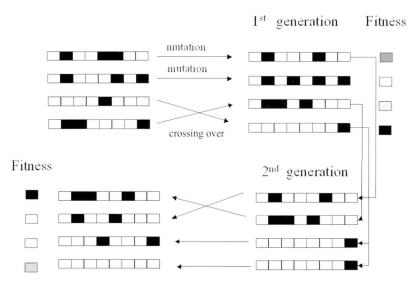

Figure 3.10 Simplified schematic diagram of a genetic algorithm. In the figure, the "genome", i.e. the encoding of the initial conformations, is made of two possible states, black and white. For a protein, this can be a bit string encoding its ϕ and ψ angles. The darker the box representing the fitness of each "individual", the higher its fitness. For proteins, this can be the value of the computed energy using, for example, a pair-wise potential. The next generation contains a number of copies of the "individuals" related to their fitness. For example, in the figure, the frequency of the fourth individual of the first generation is doubled in the second generation.

3.10
Effectiveness of Ab-initio Methods for Folding a Protein

Attempts to use molecular dynamics to fold a protein, i.e. starting from an unfolded conformation of a protein and letting it reach its global native structure, are countless. To simulate physical reality, a method must represent all the atoms of the protein and of the solvent, usually more than 10 000, use a small time step (approximately a femtosecond, as already remarked) and carry on the simulation for micro- or milliseconds. This is not yet feasible. Folding simulations have been used to study the unfolding processes of small proteins that can be accelerated by raising the simulation temperature and/or by applying external forces, and some encouraging results have been obtained. Care must still be taken when interpreting the results, however, because the short-time simulations can only sample a very limited conformational space.

It is not yet clear whether, even if one could run a very detailed simulation for a very long time, our treatment of the protein as a classical object is sufficiently accurate to be useful in folding simulations of larger, and therefore more complex, proteins. None of the available energy calculation methods and force fields yet seems able to consistently detect the lowest-energy conformation among a set of "decoy" conformations. For example the computed potential energy of a crystal structure is not necessarily found to be lower than that of related, but different, conformations of the same protein. In other words, the computed energy does not seem to be able to distinguish between correct and incorrect structures, when the latter are reasonable. We will come back to this point when we will discuss refinement of the protein structure models obtained by use of different methods.

With regard to folding, one interesting initiative that could, in principle, tell us whether the available force fields are sufficiently accurate to achieve this daunting objective is a distributed computing project, called Folding@home, that tries to span timescales thousands to millions of times longer than previously achieved. The idea consists in letting users download and run simulation software on their own computer when it is idle. Over 1 000 000 CPUs throughout the world have participated in the project in the last few years, but although simulations of several small proteins have already met with some success, the goal of folding a protein of average size seems to be still out of reach.

The success of the heuristic methods that will be described in forthcoming chapters has had the unfortunate side-effect of discouraging many scientists from pursuing the objective of folding a protein from scratch, although a successful ab-initio prediction of the structure of a small protein (48 residues) has been submitted to the sixth edition of CASP by Professor Scheraga, one of the pioneers of the field of molecular simulations (Figure 3.11). It is important that such efforts are not discontinued. We can assert that we have understood the protein-folding process only if we can reproduce it. Even if we could predict the structure of all the proteins of the universe by using the empirical knowledge-based methods that will be described in the rest of this book, we would still be lacking a deep understanding of the physics of the folding process. This is clearly unsatisfactory, both intellec-

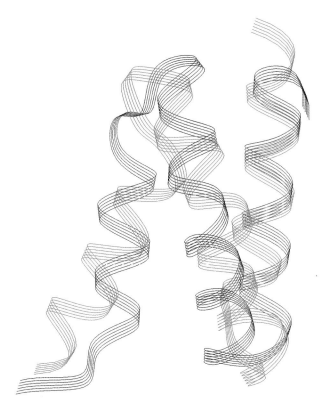

Figure 3.11 Comparison of a prediction submitted in CASP6 by Professor Scheraga (in green) and its subsequently determined experimental structure (in blue).

tually and otherwise. Understanding the process would have practically important consequences, because the principles that drive protein folding also dictate substrate and ligand binding and the induced conformational changes often associated with protein functions and explains the process of incorrect folding that is the cause of many diseases.

Suggested Reading

A seminal paper discussing the problem of modeling a protein:
C. Levinthal (**1966**) Molecular model-building by computer. Scientific American **214**, 42–52

An extremely good and interesting book on proteins physics:
A.V. Finkelstein, O. B. Ptitsyn (**2002**) Protein Physics, Academic Press, London

4
Evolutionary-based Methods for Predicting Protein Structure: Comparative Modeling

4.1
Introduction

Historically, methods for predicting protein structure are distinguished according to the relationship between the target protein(s) and proteins of known structure. "Comparative modeling" is the name given to the set of techniques that can be applied when a clear evolutionary relationship between the target and a protein of known structure can be easily detected from the sequence. "Fold-recognition" is the name given to methods that can be applied when the structure of the target protein turns out to be related to that of a protein of known structure although the relationship is difficult, or impossible, to detect from the sequences. In other words, the target protein can have a very distant relationship with a protein of known structure so that sequence-based methods are not sufficient, by themselves, to ensure they are part of the same family (homologous fold recognition) or share a fold with a known protein for reasons other than evolution (analogous fold recognition). Finally, when neither the sequence nor the structure of the target protein are similar to that of a known protein, we classify the methods as techniques for new fold prediction. The subdivision is somewhat artificial – the distinction between comparative models and homologous fold recognition is based on the ability of sequence-based methods to detect evolutionary relationships and therefore depends on the method used to detect the relationship. The distinction between analogous fold and new fold recognition, on the other hand, relies on our definition of similarity between folds and here also the demarcation line is rather fuzzy.

We will follow the classical subdivision in this book mainly for ease of reference, but the reader should not be led to believe that, in building a model that falls into one of these broad categories, she or he can safely ignore the advances and pitfalls of the others. Rather, each technique is providing information and tools that can be used throughout the range of targets of three-dimensional structure modeling. This obviously implies that one must often refer to aspects and techniques that are discussed in the context of a different

Protein Structure Prediction. Edited by Anna Tramontano
Copyright © 2006 WILEY-VCH Verlag GmbH & Co. KGaA, Weinheim
ISBN: 3-527-31167-X

methodology. To simplify matters, Figure 4.1 tries to provide a guide to structure prediction pointing to the relevant sections that contain topics related to each of the steps of the procedure.

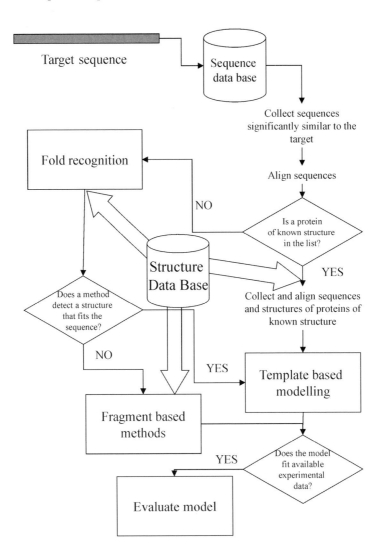

Figure 4.1 A guide to protein-structure prediction. The first step is always a search in the protein sequence database. Comparative modeling should be used when a protein of known structure sharing sequence similarity with the protein under examination is present in the database. If this is not so, fold-recognition methods should be applied and, should they fail, the user should resort to new fold or fragment-based methods. Note the central role played by the structure database in all these heuristic methods.

4.2
Theoretical Basis of Comparative Modeling

Comparative modeling is the most used method for predicting the structure of proteins. There are two reasons for this. First, the quality of models based on

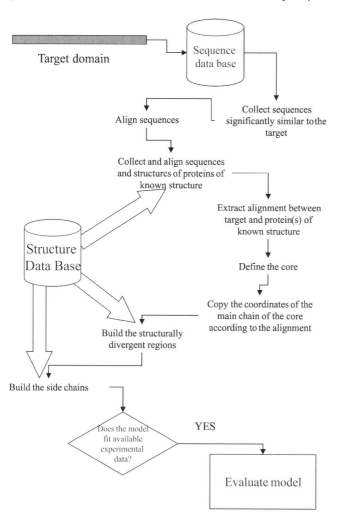

Figure 4.2 Schematic diagram of a typical comparative modeling procedure. The protein of interest should first be split into its domains. For each domain, sequences similar to the target sequences should be collected using a database search tool such as FASTA, BLAST, or PSI-BLAST. The sequences retrieved should be realigned using a multiple sequence alignment program (for example CLUSTAL or T-COFFEE). The implied alignment between the target protein and the protein(s) of known structure will form the basis of construction of the model. This can proceed by first building the main chain of the core regions, then the main chain of the structurally divergent regions, and, finally, the side-chains. The final evaluation of the model should take into account any available information on the protein of interest.

reasonably close evolutionary relationships have been shown to be more accurate, on average, that those produced with different techniques. Second, the expected reliability of the final model can be estimated a *priori*. The latter is not a trivial point, it enables a decision to be made about whether a model can be sufficiently accurate to provide the required answers for the biological problem at hand.

The fundamental concept that forms the basis of comparative modeling methods is that evolutionarily related proteins have similar conformations and, therefore, the experimental structure of a protein can be used as a starting model for that of other members of its evolutionary family. The structure of these proteins will be similar in the sense that evolutionarily related amino acids will occupy similar relative positions in homologous proteins. Consequently, two essential ingredients of comparative modeling are the ability to detect evolutionary relationships on the basis of the amino acid sequence of proteins and to deduce the correspondence between amino acids of evolutionarily related proteins that reflects the evolutionary history of the family, i.e. to derive a biologically meaningful sequence alignment. Although these two steps, by themselves, are sufficient to obtain a low-resolution model of a protein, the substitutions, insertions, and deletions that have accumulated during evolution have an effect on the structure which must be modeled, i.e. the known structure must be modified to fit the sequence of the protein under examination. The classical procedure for construction of an homology model can therefore be summarized as follows (Figure 4.2):

- given a protein of unknown structure, identify proteins of known structure that are evolutionarily related to it;
- if they exist, construct a reliable alignment, i.e. deduce the correspondence between related amino acids in the core, i.e. in regions other than those affected by insertions, deletions, and local refolding;
- assign the coordinates of the backbone atoms of the core residues of the template protein to the backbone atoms of the corresponding amino acids of the target protein according to the sequence alignment;
- model the regions outside the conserved core;
- model the position of the side-chains of the target; and
- optimize the final three-dimensional structure.

There is no reason, at least in principle, why these steps should be sequential. It is not difficult to imagine, for example, that the positioning of the side-chains of the residues of the core might be impossible or result in a very unlikely pattern if the alignment is incorrect. The process of modeling the side-chain conformations can therefore require that the original alignment is modified. Indeed this is not unusual and the alignment must often be modified during the procedure, as illustrated by the example in Figure 4.3.

Ideally, one would like to optimize the final model and not each single step of the process. The task is to find the evolutionarily related structure(s), the alignment, divergent region, and side-chain-building procedure that, taken together, optimize parameters that correlate with the accuracy of the final model. In practice, this cannot be done both because it is computationally prohibitive (too many variables

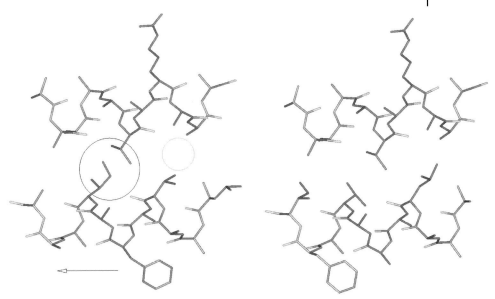

Figure 4.3 The model-building procedure might help refining the initial alignment. In the example shown here it is apparent that a shift of one residue of the sequence towards the carboxy-terminus results in a much better packing of the side-chains.

and possibilities) and because, as already mentioned, we do not have a satisfactory way of evaluating the accuracy of a model on the basis of its atomic coordinates.

Although the "classical" sequential procedure can be modified in several ways to improve each of the steps and overcome the limitations of the approach, it is still convenient to briefly discuss each step separately, because each leads to several problems and takes advantage of different techniques and methods. The sequential nature of the procedure implies that errors in one step are bound to affect all subsequent steps. Clearly, selection of the wrong template, i.e. of a protein very distant from the target when other, better, templates are available or – even worse – of a protein not evolutionarily related to the target has devastating effects on the final model; equally serious are errors in sequence alignment – it is extremely important to be very careful especially in the first part of the procedure.

4.3
Detection of Evolutionary Relationships from Sequences

The first question that arises is how to detect an evolutionary relationship between two proteins. Let us assume that the optimum sequence alignment, i.e. that which reflects the evolutionary relationship between two protein sequences, is that which minimizes the differences between them. In this hypothesis, and if we ignore for the moment insertions and deletions, the alignment can be computed exactly with

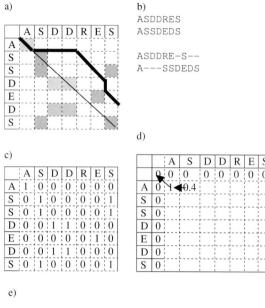

Figure 4.4 The Needleman and Wunsch alignment algorithm. A path in the matrix corresponds to an alignment. In the example, the thin line in part *a* of the figure corresponds to the first alignment shown in part *b*. The line runs diagonally and therefore corresponds to an alignment where there are no insertions or deletions. The tick line, instead, contains an horizontal line (indicating that the amino acids SDD of the first sequence do not correspond to any amino acid of the second and therefore represent an insertion in the first sequence) and two vertical lines (implying that the amino acid D and the final DS pair of the second sequence do not correspond to any amino acid in the first and is an insertion in the second sequence or, equivalently, a deletion in the first). To compute the optimum alignment we fill the cells of the matrix (part *c*) with a number representing the likelihood that the amino acid in the row is replaced by that in the column. In this example we assign 1 to identical amino acids and 0 to different ones. Part *d* shows the construction of the cumulative matrix as described in the text.

algorithms known as dynamic programming algorithms, as we will see later. Next, we can ask how likely it is that the resulting similarity between the two sequences has arisen by chance. If such a probability is very low, we can infer with some confidence that the two proteins are homologous. In other words we can measure the minimum sequence distance between two proteins and, if such a distance is small, infer with some statistical reliability that the two proteins are homologous.

Most of the algorithms for computing the optimum alignment of two protein sequences start by constructing a matrix with one sequence in the first row and the other sequence in the first column. Each cell ij of the matrix represents the alignment of residue i of the first sequence with residue j of the second, and therefore each path in the matrix corresponds to a possible alignment between the first and second sequences (Figure 4.4 a). We are interested in the path that maximizes the fraction of identical amino acids between the two sequences. If we fill the ij cell when the amino acids i and j are identical, we are looking for the path that includes the maximum number of filled cells. The path cannot run backward and can include horizontal and vertical segments: a horizontal segment of the path corresponds to an insertion in the first sequence, a vertical one to an insertion in the second, as shown in Figure 4.4 a.

We might require the algorithm to find the optimum global alignment, i.e. that starting from the first cell in the upper left corner and ending in the last, in the lower right corner, or local ones, i.e. only includes high-scoring segments (segments containing many filled cells) between the two sequences.

Insertions and deletions are rarer events than substitutions, therefore we must tell the algorithm that vertical and horizontal moves must be penalized compared with diagonal moves. It is difficult to correctly estimate the penalty that must be attributed to insertions and deletions. The values used by the different methods are determined heuristically by optimizing the alignment between proteins with known evolutionary relationships, but there is a large body of literature discussing the problem. In general, it is not a good idea to assign the same penalty to each inserted or deleted amino acid. Because insertions and deletions can be more easily accommodated in a protein structure if they occur near the solvent-exposed surface, i.e. in a limited set of positions, it is better to penalize less the continuation of a gap with respect to its initiation.

4.4
The Needleman and Wunsch Algorithm

Figure 4.4 a shows a matrix in which each row corresponds to one character of one string and each column to one character of the other. An element of the matrix is shaded if the characters corresponding to its row and column are identical. A correspondence (alignment) between the two strings is a path in the matrix, as illustrated in Figure 4.4 b. The alignment can be global, i.e., include the whole strings, or local, including only regions of them.

To find the optimal global alignment using our hypotheses, we need the correspondence between the two strings requiring the minimum number of editing operations, that is in which the number of identical corresponding characters is maximum. If we build the matrix shown in Figure 4.4 c, in which cells corresponding to pairs of identical amino acids are set to unity, we need to find the path that goes from the upper left corner to the lower right corner and includes the maximum number of cells containing "1", i.e. for which the sum of the values of the cells in the path is maximum. Here we are assuming that characters are either identical (scoring 1 in our matrix) or different (scoring 0), but the reasoning can be easily extended to instances in which cells are filled with values reflecting the similarity between the two amino acids of the row and the column, rather than their identity.

Let us build the matrix in Figure 4.4 d, called the cumulative matrix, where in each cell we write the maximum score that can be achieved by any path ending in that cell.

The column and row labeled "0" correspond to inserting or deleting at the beginning of either string. If we do not require the first characters of each string to be aligned, we can set the scores in the first row and first column to 0. Alternatively, we can add a penalty for shifting them, for example, subtracting 0.6 for each shifted character. Calculating the value in the (1, 1) cell is trivial – a path including this cell must include either the cell (0, 0) or the cell (1, 0) or the cell (0, 1). The maximum score achievable by any path ending in (1, 1) is the maximum of:

- 1, i.e. the value in (0, 0) + 1 (because the characters in the first row and the first column are identical and therefore we gain 1 by passing through the (1,1) cell);
- −0.6, i.e. the value in (0, 1) minus the penalty value for an insertion in the horizontal sequence (0.6 in the example); and
- −0.6, i.e. the value in (1, 0) – the penalty value for an insertion in the vertical sequence (0.6 in the example).

We will write this maximum value in the cell (1, 1) and store a pointer to the cell (0, 0) – the cell we used to obtain it.

When (1, 1) is filled, the same strategy can be used to calculate the values of cells (1, 2), (2, 1), and (2, 2), and so on, as shown in Figure 4.4 d. The maximum achievable score for a global alignment of the two strings must be the value in the last cell (7, 7), and this can be obtained if we passed through the cell (6, 6) which was filled using the value in (5, 5), and so on. In other words finding the best path only requires walking backward from (10, 8) to (0, 0) following the pointers.

This algorithm, which is known under the name "Needleman and Wunsch global alignment", can be easily modified to find the best local alignment by starting from the maximum value in a similarly derived cumulative matrix and working our way until we find a 0 (the Smith and Waterman algorithm). In our example these two alignments coincide. The Needleman and Wunsch algorithm guarantees that one optimum path is found. It finds the alignment of two protein sequences that maximizes their identity, or similarity, given a preassigned insertion/deletion penalty.

> Question: Is it reasonable to modify the alignment manually taking into account other information?
>
> »In comparative modeling the experimental structure of one of the proteins in the alignment is known and we know that

the structure of the other is very similar. The alignment already gives us precious information about the location of the secondary structure in both proteins and about their overall architecture, including which regions are exposed to the solvent and which are buried in the core of the proteins. By inspecting the alignment and the structure of the known protein, it is possible to manually adjust the positions where insertions and deletions are more likely to be located. For example, if the alignment algorithm has positioned a gap inside a secondary structure element, it is advisable to move it to the beginning or end of the element. This is a perfectly legitimate procedure. The alignment algorithm maximizes a predefined score and, in general, has no information about the structural features of the proteins, information that should be taken into account whenever possible.«

4.5
Substitution Matrices

The probability that a substitution is accepted in a protein is not the same for every amino acid change. It is easy to understand that the substitution of a valine with a leucine might be more frequently accepted during evolution than substitution of a glycine with a tryptophan. This can be taken into account by making our alignment matrix more sophisticated – rather than setting to unity the value of cells corresponding to identical amino acids, we can assign to each cell a value that reflects the likelihood that the amino acids in the row and the column are substituted for each other during evolution. These values are reported in substitution matrices, 21×21 tables where the twenty amino acids are in the first row and in the first column and each cell (i,j) contains a value related to the probability that the amino acid i is mutated into amino acid j. Filling the cells corresponding to pairs of identical amino acids and leaving the remainder blank corresponds to an identity substitution matrix, i.e. a substitution matrix where all cells are set to 0 except for the diagonal ones that are set to 1.

Other matrices can be based, for example, on chemical similarity between amino acids or on the minimum number of base substitutions needed to transform a triplet coding for one into one coding for the other. The most commonly used matrices, however, the PAM and BLOSUM matrices (Figure 4.5), are derived from comparisons of evolutionarily related proteins.

PAM (percent accepted mutation) is a unit introduced by Margareth Dayhoff and coworkers to quantify the amount of evolutionary change in a protein sequence. A PAM unit corresponds, on average, to 1% of accepted amino acids changes. PAM1 is a matrix calculated from alignments of sequences at 1 PAM distance from each other. Given these alignments, we compute, for each pair of amino acids i, j, the ratio $f_{ij}/f_i f_j$, where f_{ij} is the frequency with which the two

Pam 250

	A	R	N	D	C	Q	E	G	H	I	L	K	M	F	P	S	T	W	Y	V
A	2	-2	0	0	-2	0	0	1	-1	-1	-2	-1	-1	-3	1	1	1	-6	-3	0
R	-2	6	0	-1	-4	1	-1	-3	2	-2	-3	3	0	-4	0	0	-1	2	-4	-2
N	0	0	2	2	-4	1	1	0	2	-2	-3	1	-2	-3	0	1	0	-4	-2	-2
D	0	-1	2	4	-5	2	3	1	1	-2	-4	0	-3	-6	-1	0	0	-7	-4	-2
C	-2	-4	-4	-5	12	-5	-5	-3	-3	-2	-6	-5	-5	-4	-3	0	-2	-8	0	-2
Q	0	1	1	2	-5	4	2	-1	3	-2	-2	1	-1	-5	0	-1	-1	-5	-4	-2
E	0	-1	1	3	-5	2	4	0	1	-2	-3	0	-2	-5	-1	0	0	-7	-4	-2
G	1	-3	0	1	-3	-1	0	5	-2	-3	-4	-2	-3	-5	0	1	0	-7	-5	-1
H	-1	2	2	1	-3	3	1	-2	6	-2	-2	0	-2	-2	0	-1	-1	-3	0	-2
I	-1	-2	-2	-2	-2	-2	-2	-3	-2	5	2	-2	2	1	-2	-1	0	-5	-1	4
L	-2	-3	-3	-4	-6	-2	-3	-4	-2	2	6	-3	4	2	-3	-3	-2	-2	-1	2
K	-1	3	1	0	-5	1	0	-2	0	-2	-3	5	0	-5	-1	0	0	-3	-4	-2
M	-1	0	-2	-3	-5	-1	-2	-3	-2	2	4	0	6	0	-2	-2	-1	-4	-2	2
F	-3	-4	-3	-6	-4	-5	-5	-5	-2	1	2	-5	0	9	-5	-3	-3	0	7	-1
P	1	0	0	-1	-3	0	-1	0	0	-2	-3	-1	-2	-5	6	1	0	-6	-5	-1
S	1	0	1	0	0	-1	0	1	-1	-1	-3	0	-2	-3	1	2	1	-2	-3	-1
T	1	-1	0	0	-2	-1	0	0	-1	0	-2	0	-1	-3	0	1	3	-5	-3	0
W	-6	2	-4	-7	-8	-5	-7	-7	-3	-5	-2	-3	-4	0	-6	-2	-5	17	0	-6
Y	-3	-4	-2	-4	0	-4	-4	-5	0	-1	-1	-4	-2	7	-5	-3	-3	0	10	-2
V	0	-2	-2	-2	-2	-2	-2	-1	-2	4	2	-2	2	-1	-1	-1	0	-6	-2	4

Blosum 62

	A	R	N	D	C	Q	E	G	H	I	L	K	M	F	P	S	T	W	Y	V
A	4	-1	-2	-2	0	-1	-1	0	-2	-1	-1	-1	-1	-2	-1	1	0	-3	-2	0
R	-1	5	0	-2	-3	1	0	-2	0	-3	-2	2	-1	-3	-2	-1	-1	-3	-2	-3
N	-2	0	6	1	-3	0	0	0	1	-3	-3	0	-2	-3	-2	1	0	-4	-2	-3
D	-2	-2	1	6	-3	0	2	-1	-1	-3	-4	-1	-3	-3	-1	0	-1	-4	-3	-3
C	0	-3	-3	-3	9	-3	-4	-3	-3	-1	-1	-3	-1	-2	-3	-1	-1	-2	-2	-1
Q	-1	1	0	0	-3	5	2	-2	0	-3	-2	1	0	-3	-1	0	-1	-2	-1	-2
E	-1	0	0	2	-4	2	5	-2	0	-3	-3	1	-2	-3	-1	0	-1	-3	-2	-2
G	0	-2	0	-1	-3	-2	-2	6	-2	-4	-4	-2	-3	-3	-2	0	-2	-2	-3	-3
H	-2	0	1	-1	-3	0	0	-2	8	-3	-3	-1	-2	-1	-2	-1	-2	-2	2	-3
I	-1	-3	-3	-3	-1	-3	-3	-4	-3	4	2	-3	1	0	-3	-2	-1	-3	-1	3
L	-1	-2	-3	-4	-1	-2	-3	-4	-3	2	4	-2	2	0	-3	-2	-1	-2	-1	1
K	-1	2	0	-1	-3	1	1	-2	-1	-3	-2	5	-1	-3	-1	0	-1	-3	-2	-2
M	-1	-1	-2	-3	-1	0	-2	-3	-2	1	2	-1	5	0	-2	-1	-1	-1	-1	1
F	-2	-3	-3	-3	-2	-3	-3	-3	-1	0	0	-3	0	6	-4	-2	-2	1	3	-1
P	-1	-2	-2	-1	-3	-1	-1	-2	-2	-3	-3	-1	-2	-4	7	-1	-1	-4	-3	-2
S	1	-1	1	0	-1	0	0	0	-1	-2	-2	0	-1	-2	-1	4	1	-3	-2	-2
T	0	-1	0	-1	-1	-1	-1	-2	-2	-1	-1	-1	-1	-2	-1	1	5	-2	-2	0
W	-3	-3	-4	-4	-2	-2	-3	-2	-2	-3	-2	-3	-1	1	-4	-3	-2	11	2	-3
Y	-2	-2	-2	-3	-2	-1	-2	-3	2	-1	-1	-2	-1	3	-3	-2	-2	2	7	-1
V	0	-3	-3	-3	-1	-2	-2	-3	-3	3	1	-2	1	-1	-2	0	-3	-1	4	

Figure 4.5 The PAM250 (part *a*) and BLOSUM62 (part *b*) substitution matrices. The values corresponding to pairs of amino acids can be used to fill the alignment matrix (part *c* of Figure 4.4).

amino acids are found in corresponding positions in the alignments, i.e. have replaced each other during evolution, and f_i and f_j are the frequencies with which the amino acids i and j occur in the sequences (their product is an estimate of the probability that the two amino acids are found in corresponding positions by chance alone, given the composition of the sequences in the alignment). The ratio $f_{ij}/f_i f_j$ is an estimate of the likelihood that the amino acids i and j are substituted by each other during evolution. Similarity matrices usually report the logarithm to base 2 of these numbers. PAM2 is calculated by multiplying PAM1 by PAM1, PAM3 by multiplying PAM2 by PAM1, and so on (Figure 4.5 a). PAM matrices are based on global alignments of closely related proteins. The higher the number of the matrix, the more suitable it is for aligning distantly related sequences.

The BLOSUM (*blocks substitution*) matrices are instead derived using local alignments of very conserved regions in homologous proteins. They also come as a series of matrices. A BLOSUM-N matrix is derived from alignments such that all sequences sharing more than $N\%$ identity with any other sequence in the alignment are averaged and represented as a single sequence (Figure 4.5 b). In contrast with PAM, here a larger number indicates a matrix more suitable for aligning more closely related sequences.

Because substitution matrices, schemes for gap penalty scores, and alignment algorithms are extensively described in many books and articles, we will not go into further detail here and will directly ask the next question – given an alignment score (i.e. the value reported by our alignment algorithm) how likely it is that it reflects a true evolutionary relationship rather than a random similarity of the two amino acid sequences that are, after all, both composed from the same twenty amino acids? In other words, could our alignment score have been obtained when aligning two independent, not evolutionarily related, protein sequences with the same length and composition? If the score obtained by aligning two sequences is significantly higher than the distribution of scores obtained by aligning sets of unrelated sequences, we can say with some confidence that the two aligned sequences are evolutionarily related, but obtaining the "background" distribution of scores for unrelated sequences is not straightforward, because we do not have a validated set of sequences which are definitely unrelated to the ones under examination. One way to approach the problem is to reshuffle our sequences many times, thus obtaining pairs of synthetic sequences with the same composition as our original sequences, but random and therefore not evolutionarily related, align them in pairs and collect the scores for each alignment. The distribution of these scores represents the distribution expected for the scores of alignments between unrelated sequences. We can statistically evaluate whether the score obtained in the real alignment is likely to belong to this random distribution, in which case we assume it is not significant. Indeed programs such as BLAST, PSI-BLAST, and FASTA designed to search a data base for protein sharing a significant sequence similarity to a query protein use this principle to compute the reported probability values, as will be discussed next.

4.6
Template(s) Identification Part I

Step 1:
Given a Protein of Unknown Structure, Identify Proteins of Known Structure that are Evolutionarily Related to it

The task is to use the sequence information on the target protein to detect evolutionary relationships with proteins of known structure. The basic scheme consists in aligning the sequence of the protein of interest with every protein of known structure, compute the sequence identity or similarity and compare the observed value with that expected by chance alone. This is, roughly, what programs such as FASTA or BLAST do – given a "query" protein sequence, they align it with every protein in the data base and report the score of the alignment together with a value related to the probability that the observed similarity is statistically significant. In practice, because of the size of the data base, both programs use approximations to avoid aligning the target sequence with every single sequence. For example, they discard proteins not having at least a short peptide (two or three residues) identical with one of the query protein, compute an approximate score for the alignment of the target with the remaining ones, sort them accordingly, and perform a full fledged alignment only on the subset of sequences likely to be homologous to the query. The details of the approximations used will not be discussed here, both because they are reported in many books, manuals, and articles, and because they are subject to changes in different versions of the tools.

The most relevant issue, as far as the template selection step of comparative modeling is concerned, is evaluation of the score, i.e. how do we decide whether or not a given match is indicative of a structural similarity or, which is equivalent, of an evolutionary relationship? If we are searching a data base to attempt the functional assignment of a protein, we must distinguish between orthologous and paralogous relationships. In modeling this is less of an issue, because both orthologous and paralogous proteins are expected to share a similar structure. We want to make sure that the observed similarity is really indicative of an evolutionary relationship, however.

Both programs use a similar strategy – they compare the observed score between the query protein and each of the database proteins with a random distribution of scores and evaluate how likely it is that the observed similarity belongs to this reference distribution and is, therefore, not significant. With FASTA, the reference distribution of scores is obtained by randomly shuffling the sequence of the query sequence and repeating the search many times on subsets of the original database. BLAST, instead, computes the expected distribution of random scores using as query sequences a set of random sequences with the average composition and length of the database sequences. This implies that BLAST is faster because it does not need to reconstruct the random distribution for each database search, but also that the distribution obtained is based on the assumption that the composition of the query sequence does not differ substantially from the average composition of the proteins in the database. This might not be true – the query sequence can be biased and be particularly rich in some amino acids, invalidating the statistics. Because of this, the first step in BLAST is to "mask", i.e. not to use in the scoring, those parts of the input sequence that deviate from the composition of the reference distribution. It is important to ensure the masking option of BLAST is turned on; the program output will indicate which regions have been masked and the user can critically evaluate whether they have a relevant effect on the statistics of the results.

> *Question: Why is it important to take into account the composition of the query protein in evaluating the significance of the score?*
>
> »The assumption that the composition of the query sequence is similar to that of the sequences used to derive the random distribution is very important for obtaining meaningful results. Let us assume that our sequence is full of, say, prolines. There will be a high chance that the prolines in our sequence match other prolines in the sequences of the database, even if they are evolutionarily unrelated to the query, and these matches will contribute to increase the score of the alignment. The average score of the random distribution, derived using sequences with a more regular distribution of amino acids, and not containing many matching prolines, will be lower than that expected for the alignment of two random sequences containing many prolines. On average, the random scores will be lower than those obtained by our query sequence with unrelated, but proline-rich, sequences and this might lead us to incorrectly assume that the similarity with the latter is statistically significant.«

How do we measure likelihood in a database search, be it the one from BLAST or FASTA? As we said, we need to compare the score with the expected random distribution and calculate the probability that the observed score belongs to the distribution. For example, if the random distribution were Gaussian, the probability of obtaining a value that is one (three) standard deviation(s) above the mean is

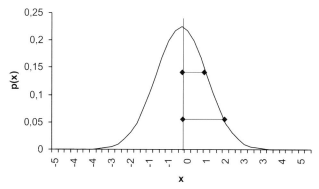

Figure 4.6 A Gaussian distribution with mean = 0 and $\sigma = 1$. The two segments correspond to one and two standard deviations.

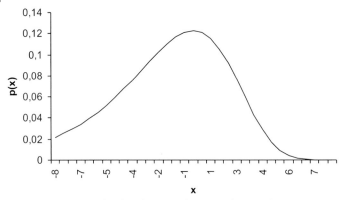

Figure 4.7 Extreme value distribution with μ = 0 and β = 3. This is the expected distribution for the alignment scores of unrelated sequences.

6% (0.15%). If we obtain a score which is, say, three standard deviations above the mean of the random distribution we can expect to find it by chance alone once every seven hundreds times when aligning unrelated sequences. If we are searching in a database containing two million sequences, we expect to find a few thousand sequences with such a score or higher (Figure 4.6).

The expected distribution of scores of random alignments is not Gaussian. It has been shown computationally that it resembles another type of distribution called "extreme value distribution" (Figure 4.7) often used to model the smallest or largest value among a large set of independent, identically distributed random values representing measurements or observations.

This simply means that the probability of observing a value x, following an extreme value distribution, is:

$$p(x) = 1/\beta \cdot e^{\frac{x-\mu}{\beta}} e^{-e^{\frac{x-\mu}{\beta}}}$$

where β is positive. μ is the location parameter (related to the "position of the distribution" with respect to the x axis) and is estimated by fitting the distribution to the data. The output of a database search with BLAST or FASTA will report the E value, i.e. the expected number of alignments obtaining that score by chance alone, according to the theory of extreme value distribution. The E value for the ungapped alignment of two unrelated sequences of lengths m and n having a score S is:

$$E(S) = Kmne^{-\lambda S}$$

where K and λ are terms that depend on which similarity matrix we use for the amino acids and on the composition of the two sequences. It can also be shown that the cumulative probability of obtaining an alignment with score $S^* \geq S$ is given by:

$$p(S^* \geq S) = 1 - e^{-E(S)}$$

When we compare a query sequence of length m with a database of sequences, we are performing several comparisons with the same m but different n. The approximation used by BLAST is to consider the data base as a single very long sequence and to use for n the value of the total length of the database. Therefore, if the score for the comparison of a query sequence m residues long with one of the database sequences is S^*, the probability that the score is higher than expected by chance is:

$$p = 1 - e^{-KmNS^*}$$

where N is the total length of the database, i.e. the number of residues in the database. This theory is strictly valid for alignments without gaps (insertions and deletions), but it is difficult to prove when gaps are included in alignments; computational simulations have shown, however, that it applies fairly well to gapped alignments also. Notice that the normalized score S' (the bit score) is defined as:

$$S' = (\lambda S - \ln K)/\ln 2$$

The expected value, E, now becomes:

$$E = mn2^{-S'}$$

and this can be used to compare the results of different database searches, because it is independent of K and λ.

When we run a database search program, each match reports the score, or the bit score, the E value and the p value. Clearly we are confident in matches for which the E value is very low and the p value very high. When is a match significant or, in other words, what is the right E value or p value threshold that we should use to effectively discriminate between biologically meaningful and random matches? This is one of the most difficult questions in protein bioinformatics, and to understand how to deal with the issue, we need to introduce the concept of sensitivity and specificity. Let us assume we have a protein and a database of sequences some of which are known to be homologous to the query sequence and some known to be unrelated. The set of sequences can be obtained by only including proteins with known three-dimensional structure for which the evolutionary relationship is easier to assess. Now we run a database search and obtain a list of proteins, each with an associated E value. Ideally, we would like all the evolutionarily related proteins to have E values lower than that of the unrelated ones, but this is unlikely to happen. What we can do is to examine the E values and, for each of them, ask ourselves:

If this value were selected as a threshold for separating related and unrelated proteins, which fraction of the related proteins would be missed and which fraction correctly identified (i.e. how many truly homologous proteins would have an E value lower than the threshold and how many an E value higher than the threshold)?

We call the former "false negatives" (FN) and the latter "true positives" (TP). We then ask ourselves which fraction of unrelated proteins would be mistakenly labeled as evolutionarily related (with an *E* value higher than the threshold) and which fraction of unrelated ones would, correctly, have an *E* value lower than the threshold? These are, respectively the "false positives" (FP) and "true negative" (TN). We define (see Figure 4.8):

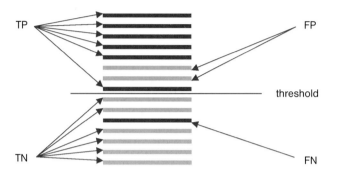

Sensitivity = 6/7 = 0.86
Specificity = 6/8 = 0.75

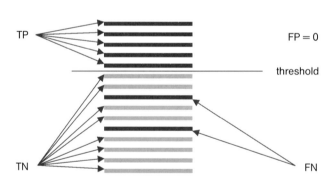

Sensitivity = 5/7 = 0.71
Specificity = 8/8 = 1.00

Figure 4.8 Examples of sensitivity and specificity values for a database search method. In the figure, dark and light segments, respectively, represent proteins homologous and unrelated to the query sequence. If we select the threshold as shown in the top part of the figure, two unrelated sequences will be labeled as "homologous" and one homologous one as "unrelated". A more stringent threshold (bottom), will eliminate false positives, but will increase the number of false negatives.

$$\text{Sensitivity} = TP/(TP + FN)$$

$$\text{Specificity} = TN/(TN + FP)$$

It is important to understand the meaning of these two values and, as always when trying to grasp the meaning of an equation, it is convenient to look at extreme cases. If our E value is very high and we accept every match as significant, we would not miss any TP, no related sequences would escape our attention, but we would also assume that every sequence in the database is related to the query, therefore we would have TN = 0, FN = 0, and, therefore, Sensitivity = 1, Specificity = 0. Our selection would be very sensitive, i.e. not miss anything, but very aspecific, because it would not specifically distinguish between related and unrelated sequences. In the opposite case, where our E value threshold is set to a very low value and every alignment turns out to have an E value higher than the threshold, we would have TP = 0, FP = 0, and Sensitivity = 0 and Specificity = 1. In other words we would never mistakenly assume that two sequences are evolutionary related, but we would not detect most of the true relationships. In summary, a less stringent threshold will increase the fraction of correctly identified true positive, but will also increase the fraction of false positives. If we plot these two values, the true positive and the false positive fraction, against each other, we obtain what is called a ROC (receiving operator curve). The details of the curve clearly depend on the choice of the query sequences, the database, and the settings. An example of a set of ROC curves, with different stringency values, is shown in Figure 4.9.

The FASTA and BLAST strategies are not the only options available for database searching. At least two other general methods are commonly used and, often, are essential for detecting evolutionary relationships – these are profile-based methods and hidden Markov models. They will be discussed after the section describing multiple sequence alignments.

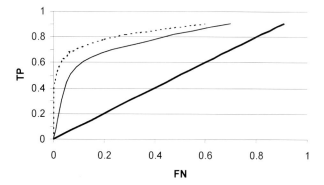

Figure 4.9 Examples of ROC curves. The tick line corresponds to a worthless method, unable to discriminate between positives and negatives. The method represented by the dotted curve is better than that represented by the continuous line: it detects more true positives when finding the same number of false negatives.

4.7
The Problem of Domains

One problem often encountered in database searches is that the query protein might be formed by domains and the detected similarity with a protein sequence in the database can be limited to one of the domains. As already discussed, the significance of the score depends on the length of the query sequence, and assumes that the detected match spans the whole sequence. The database search should be performed separately for each domain of the target protein, but the problem of how to detect the boundary of the domains in a protein from its sequence is still open and there is no clear solution. We can only provide a few practical suggestions.

The size of domains in proteins is usually in the range of 100–200 residues; if, therefore, the sequence of the query protein is much longer than this, it is very likely that the protein is multi-domain. We can try and use one of the methods developed to detect protein domain boundaries. Some are based simply on the expected size of a domain; some take into account the sequence of amino acids, trying to detect linkers between domains; yet others are based on more sophisticated techniques. Although none can guarantee perfect accuracy, their combination can help with the decision whether the sequence must be split, and approximately where.

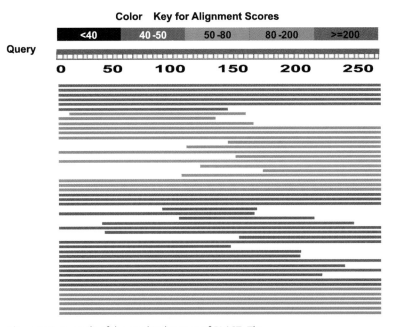

Figure 4.10 Example of the graphical output of BLAST. The example shown suggests that the query protein is formed by two domains, one spanning from the beginning to approximately residue 150, the other from approximately residue 150 to the end of the protein.

Another strategy is simply to split a large sequence into overlapping fragments of the size expected for a domain. We can run our database search using fragment 1–200, then fragment 50–250 and so on.

If the sequence is not extremely long, we can run a first database search using the complete sequence. If a limited region of our protein sequence matches a complete protein, the region probably corresponds to a domain (Figure 4.10) and we should repeat the search only using the sequence of the putative domain. The latter is important in obtaining a significant score. Next, we should search the database with the remaining part of the sequence. A significant match in this latter region might not have shown up in the first search, because both the E and p values depend upon the length of the match.

Finally, it is always a good idea, when we find a match, to "invert the search". In other words, if, in a database search, a significant match is found between protein A and protein B, we should now repeat the search using protein B as a query and verify that protein A is found with a similarly significant score. This procedure protects us from false positive matches arising as a result of peculiar characteristics of our query sequence. If the inverted search does not find the original sequence, it is likely that our search results are not biologically significant and more careful inspection of the properties of the query protein is required.

4.8
Alignment

Step 2:
Construct a Reliable Alignment of the core, i.e. Deduce the Correspondence Between Related Amino Acids in Regions Other than Those Affected by Insertions, Deletions and Local Refolding

We have already outlined the algorithm, the scoring system, and the gap penalty schemes used in alignment, but there are a few more aspects that must be discussed. First, let us recall that any alignment algorithm maximizes the conservation or similarity of paired amino acids. This is because we assume that evolution has preserved those amino acids that are essential for the function and structure of the protein and, therefore, their conservation can guide us through the process of alignment. The conservation of some amino acids might, however, be just because of chance and, when comparing two sequences, we have no way of knowing which pairing of similar amino acids should be weighted more than others because they reflect a genuine evolutionary relationship. One way to overcome this problem, at least partially, is to resort to multiple sequence alignments, i.e. to align together as many sequences as possible of proteins of the family rather than just the target and template sequences. In this way, the amino acids conserved by chance will be different for each pair of sequences, whereas those conserved because of a structural and/or functional constraint will be common to the whole family, as shown in Figure 4.11.

```
Prot1  ILSILHTYSSLNHVYKCQNK.EQFVEVMASALTGYLHTIS..SENLLDAVYSFCLMNYFPLAPFNQLLQKDII
Prot2  IVSILHVYSSLNHVHKIHN..REFLEALASALTGCLHHIS..SESLLNAVHSFCMMNYFPLAPINQLIKENII
Prot3  ISALMEPFGKLNYL..PPNA.SALFRKLENVLFTHFNYFP..PKSLLKLLHSCSLNECHPVNFLAKIFKPLFL
Prot4  IAELIEPFGKLNYV..PPNA.PALFRKVENVLCARLHHFP..PKMLLRLLHSCALIERHPVNFMSKLFSPFFL
Prot5  VQKLVLPFGRLNYL..PLE..QQFMPCLERILARE.AGVA..PLATVNILMSLCQLRCLPFRALHFVFSPGFI
Prot6  VAKILWSFGTLNYK..PPNA.EEFYSSLINEIHRKMPEFNQYPEHLPTCLLGLAFSEYFPVELIDFALSPGFV
Prot7  IPAIIRPFSVLNYD..PPQR.DEFLGTCVQHLNSYLGILD..PFILVFLGFSLATLEYFPEDLLKAIFNIKFL
Prot8  VCSVLLAFARLNFH..PEQEEDQFFSMVHEKLDPVLGSLE..PALQVDLVWALCVLQHVHETELHTVLHPGLH
Prot9  LCSVLLAFARLNFH..PDQE.DQFFSLVHEKLGSELPGLE..PALQVDLVWALCVLQQAREAELQAVLHPEFH
```

Figure 4.11 A multiple sequence alignment. Note that completely conserved amino acids are easier to detect when more sequences are considered.

This strategy is always beneficial, but it becomes essential when we are attempting to align two distantly related sequences.

The algorithm that we described before for aligning two sequences cannot be extended to many sequences, because it becomes computationally too expensive; multiple alignment methods are, therefore, usually built heuristically. In practice, one first aligns two sequences, then a third to the first two, than a fourth to the first three and so on, with an incremental approach. The alignment algorithm can be extended to align a sequence to an alignment, or an alignment to an alignment (Figure 4.12). Each of the two input sequences or alignments is treated as single sequences but the score at aligned positions is calculated as the average similarity matrix score of all the residues in one alignment relative to all those in the other alignments.

Calculation of the final scoring of a multiple alignment is a more complex problem. One rather unsatisfactory, and yet commonly used, method is to add

Alignment of

PTLRS
LTTRS with PTLR:

	P L	T T	L T	R R	S S
P	(Score (P,P) + score (L,P))/2	(Score (T,P) + score (T,P))/2	...	...	...
T	(Score (P,T) + score (L,T))/2	(Score (T,T) + score (T,T))/2	...	...	...
L	(Score (P,L) + score (L,L))/2	(Score (T,L) + score (T,L))/2	...	...	...
R	(Score (P,R) + score (L,R))/2	(Score (T,R) + score (T,R))/2	...	...	...

	P L	T T	L T	R R	S S
P	(7-3)/2=2	...	...	...	...
T	(-1-1)/2=-1	...	...	...	...
L	(-3+4)/2=0.5	...	...	...	...
R	(-2-2)/2=-2	...	...	...	...

Figure 4.12 The method for aligning a sequence to an alignment. The alignment is written in the first rows of a matrix and the sequence in the first column. Each cell contains the average between the score of each amino acid of the alignment with the corresponding amino acid of the sequence. The alignment strategy, once the matrix is filled, is identical with that outlined in Figure 4.4.

Sum of pairs score for the alignment :
PTLRS
LTTRS
PTLRT

P	T	L	R	S
L	T	T	R	S
P	T	L	R	T
Score (P,L) + Score (P,P) + Score (L,P) = -3+7-3=1	Score (T,T) + Score (T,T) + Score (T,T) = 5+5+5=15	Score (L,T) + Score (L,L) + Score (T,L) = -1+4-1=2	Score (R,R) + Score (R,R) + Score (R,R) = 5+5+5=15	Score (S,S) + Score (S,T) + Score (S,T) = 4+1+1=6
		Score = 39		

Figure 4.13 The score of a multiple alignment can be computed by averaging the scores of each column, as shown in the figure.

the score of each amino acid pair in a column of the multiple alignment and to add the column scores to obtain the alignment score (Figure 4.13). There are several problems with this approach, one of which can be illustrated in Figure 4.14, in which a tree representing the similarity between members of a protein family is shown. Let us consider a multiple alignment including sequences A, B, and C. The sum of scores, for each column, will add the score of the amino acids of A and B, A and C, and B and C. We can think of it as measuring the distance along the path connecting the various nodes, and it is easy to see that the edge indicated by the thicker line in the figure is counted more than once.

Two widely used methods for multiple sequence alignment are CLUSTAL and T-COFFEE. CLUSTAL uses exhaustive pairwise alignments between all the sequen-

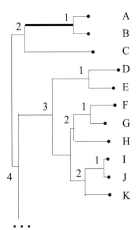

Figure 4.14 A tree constructed on the basis of the sequence similarity among several proteins (indicated by the filled circles). The numbers indicate the order in which the sequences should be iteratively aligned by use of the method described in Figure 4.12, starting from the leaves and proceeding toward the root of the tree.

ces to produce a measure of sequence similarity from which it derives a joining order. This joining order corresponds to a tree that is used to produce the multiple sequence alignment. It should be noted that this tree is not an evolutionary tree. After the joining order has been determined, CLUSTAL aligns pairs of sequences, or pairs of alignments, or one sequence and one alignment starting from the leaves of the tree, so that the most similar sequences are aligned first and most dissimilar ones are added last (Figure 4.14). In CLUSTAL, the matrix used for each alignment is selected according to the sequence distance between the set of sequences to be aligned and the gap penalties depend on the amino acids observed in the column (for example, the presence of hydrophilic or flexible residues in a column reduces the penalty for a gap in that position). The gap penalty is increased for columns that do not contain gaps, if gaps are present nearby in the alignment.

T-COFFEE computes approximate global and local pairwise comparisons of the input sequences and, from these, compiles a list of all pairs of residues observed in at least one of the local alignments, weighted by a factor that depends on the quality of the local alignment around each residue and on how many times that particular pair has been observed. The rationale of this choice is that reliable local alignments are expected to be produced more often by different methods than unreliable ones. These weights, after some more manipulation to ensure consistency, are used as similarity values between the residues for building the multiple sequence alignment.

A multiple sequence alignment can be used to construct profiles, i.e. probability tables that tell us which is the probability of observing each amino acid in each position of a family of proteins. Given a multiple alignment, we can compute the number of occurrences of each amino acid at each position divided by the number of sequences. If the number of sequences is sufficiently high, the resulting table can be seen as a probability table, reporting, for each position, the probability that a given amino acid is present.

Given a newly aligned sequence, we can calculate the probability that the sequence "fits the profile". For each position of the alignment we multiply the probability value corresponding to the amino acid of the new sequence. Because multiplying probabilities is not optimum (they are always less than 1 and their product rapidly becomes very small), the profile is expressed in terms of logarithms of the frequency values. The frequency table can contain zeroes, and the logarithm of 0 is infinite. To avoid this problem, we add 1 to each amino acid count (method of pseudo-counts). With this strategy, profiles can be used to evaluate the probability that a sequence belongs to the family of proteins used to build the profile.

Another, related, but more sophisticated method is to use a hidden Markov model (HMM). This is a way of representing a multiple sequence alignment in terms of "transition" probabilities. We can use an existing multiple alignment to evaluate, for each position, the probability that it is followed by an insertion, a deletion, or a match, and, for the last, the probability it contains each of the twenty amino acids. A hidden Markov model is a representation of the alignment (and therefore of the family of proteins) in probabilistic terms and it can be used to estimate the probability that a new sequence matches the "family model". The very simple HMM shown in Figure 4.15 is derived from the multiple sequence alignment

a)

Multiple sequence alignment:

```
A C C - E
E C E - A
A C E A A
C - E - E
```

Counts	Begin-1	1-2	2-3	3-4	4-5	5-end
Match-match	4+1	3+1	3+1	1+1	1+1	4+1
Match-delete	0+1	0+1	0+1	3+1	0+1	0+1
Match-insert	0+1	1+1	0+1	0+1	0+1	0+1
Insert-match	0+1	0+1	0+1	0+1	0+1	0+1
Insert-delete	0+1	0+1	0+1	0+1	0+1	0+1
Insert-insert	0+1	0+1	0+1	0+1	0+1	0+1
Delete-match	0+1	0+1	1+1	0+1	3+1	0+1
Frequencies						
Match-match	0.45	0.36	0.36	0.18	0.18	0.45
Match-delete	0.09	0.09	0.09	0.36	0.09	0.09
Match-insert	0.09	0.18	0.09	0.09	0.09	0.09
Insert-match	0.09	0.09	0.09	0.09	0.09	0.09
Insert-delete	0.09	0.09	0.09	0.09	0.09	0.09
Insert-insert	0.09	0.09	0.09	0.09	0.09	0.09
Delete-match	0.09	0.09	0.18	0.09	0.36	0.09

Counts						
A		2+1	0+1	0+1	1+1	2+1
C		1+1	3+1	1+1	0+1	1+1
E		1+1	0+1	3+1	0+1	2+1
Frequencies						
A		0.43	0.17	0.14	0.5	0.37
C		0.29	0.67	0.28	0.25	0.25
E		0.29	0.17	0.57	0.25	0.37

b)

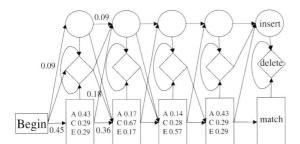

Figure 4.15 Construction of a hidden Markov model. Given a multiple sequence alignment, we first count how many times each transition (Match–match, Match–delete, Match–insert, Insert–match, Insert–delete, Insert–insert, Delete–match) occurs, and add 1 to each. The counts are then transformed into frequencies and used to construct the scheme shown in part b. For each match, we also count how many times each amino acid is observed (always adding 1 to the counts) and compute the frequencies. The scheme represents the hidden Markov model of the family of aligned proteins and can be used to calculate the probability that a new sequence belongs to the family, i.e. is generated by the HMM.

shown in the same figure, assuming there exist only three amino acids. In practice, for each position of the alignment we compute the probability that it is followed by a match, an insert, or a delete state, by counting the occurrences of matches, insertions, and deletions in the input alignment (and adding 1 for the pseudo-count). For the match states, we also compute the probability that each of the amino acids is observed in that position. Figure 4.15 shows a commonly used graphical representation of an HMM. If now we have a new sequence, we can compute the probability of the path in the HMM corresponding to it, using algorithms similar to those used for sequence alignments. In this way we can estimate the probability that the new sequence belongs to the family used to construct the HMM.

4.9
Template(s) Identification Part II

Both profiles and HMM can be used to evaluate the probability that a new sequence belongs to the family that generated them. It is only natural to use them to improve the sensitivity of data base searches. A very popular program, called PSI-BLAST is, indeed, based on profile searches. The program first runs a database search using BLAST; it then aligns the sequences that have a significant score, builds a profile, and repeats the search, this time using the profile as query. The new search will, hopefully, identify other sequences belonging to the family that can be aligned to them, so that a new profile can be built and the procedure repeated. In principle, the cycle should be repeated until no new sequences are collected. In practice, there is a significant risk that some false positive proteins slip into the list used to build the profile, making it progressively less specific for the family of interest. This is obviously because a score that appears statistically significant is not necessarily indicative of a true evolutionary relationship, as we have discussed. With PSI-BLAST the problem is aggravated, because statistical evaluation of the significance of the score is much more difficult when multiple alignments and profiles are involved.

As a rule of thumb, one should run three to five iterations of PSI-BLAST. The program allows the user to look at the profile after each iteration and its inspection can help understanding whether it is still representative of the family. The profile can be used to extract an optimum sequence with the highest scoring amino acid at each position. If this optimum sequence is very different from the query sequence, it is very likely that unrelated sequences have been added to our profile, "poisoning" it.

In specific cases, other factors can be taken into account for evaluating the appropriateness of the profile at each iteration step. For example, if proteins of known structure are found among those selected by PSI-BLAST, the user can verify that they have a similar structure and that the alignment of their sequences implied by the profile corresponds to a good structural superposition. If it is known that some amino acids play an essential role in the function or the structure of the family, they should be conserved in each of the sequences and, obviously, included in the optimum sequence.

Although comparison of a new sequence with the HMM of a family is a rather rapid procedure, construction of an HMM is rather computationally expensive. For this reason, one usually compares the query sequence with pre-computed HMM representing different families. These are stored in publicly accessible databases such as PFAM that contains thousands of HMM covering almost three quarters of the known protein sequences.

Question: Can I use the alignment provided by BLAST, FASTA or PSI-BLAST as it is to build a comparative model?

»Although database searching methods have very much improved our ability to detect evolutionary relationships and are continuously being optimized, it is important to understand that the pairwise sequence alignment that they produce is not necessarily the most accurate and that, once a database search has been used to collect the sequences of the family, these should be re-aligned using more accurate methods such as CLUSTAL or T-COFFEE.«

Question: When searching for homologous proteins of known structure, should I only search the database containing the sequences of proteins of known structure?

»Even if we are looking for a similarity between the sequence of our query protein and that of a protein of known structure, we should not limit ourselves to searching the sequences of proteins contained in the PDB database. Proteins of unknown structure are extremely useful for building better profiles and alignments. We should search the entire database, build the alignment of all the members of the family, and only afterwards extract, from this alignment, the alignment of the query sequence with the proteins of known structure.«

4.10
Building the Main Chain of the Core

Step 3:
Assign the Coordinates of the Backbone Atoms of the Core Residues of the Template Protein to the Backbone Atoms of the Corresponding Amino Acids of the Target Protein According to the Sequence Alignment

How do we identify the core or, equivalently, how do we single out those regions that are likely to have preserved their structure during evolution?

If all we have are the sequences of the target and template proteins and the structure of the latter, the problem is complex. Certainly, regions surrounding insertions and deletions have changed their structure somehow, but what about the others? One can reasonably safely assume that the main elements of secondary structure and residues buried in the protein structure have retained their conformation. But how do we treat exposed loops without insertions and deletions or small domains that are peripheral to the protein structure? Should we assume they are conserved and approximate their structure using the coordinates of the corresponding regions of the template, or try to rebuild them with one of the methods that we will describe later?

Let us recall that for proteins that are closely related, the core is expected to be formed by almost the complete structure, but its extent can decrease rapidly at increasing evolutionary distance (Figure 1.24). If a multiple sequence alignment of the protein family is available, it is usually easier to identify the conserved core. It will correspond to those regions that do not contain insertion or deletions and are well conserved in most of the members of the family. Even more favorable is when we have available the structure of more than one protein of the family. Structural superposition of these proteins will highlight regions that are more prone to change their structure during evolution and these should not be regarded as part of the conserved core. The problem of optimally superposing two or more protein structures does not have a unique solution, but in this case the problem is less severe, because our objective is only to highlight the structurally variable regions and not to obtain precise residue-to-residue correspondence between the proteins.

Another possibility is to see how stable our alignment is, i.e. to rerun the alignment procedure modifying the substitution matrix and, to a limited extent, the gap penalties, to highlight the regions where the alignment is very dependent upon our heuristic variables. When the core has been identified one could use the classical procedure of simply assigning the coordinates of the backbone of the template to the target according to the alignment. In practice, as we will see, more sophisticated methods give better results. In general, when we have at our disposal more than one structure from related proteins we can choose to use different models for different regions, according to the local sequence similarity or take into account all of the homologous structures. The SwissModel server uses this latter approach – after the templates have been identified, their structures are superimposed and the coordinates of well fitting atoms between the various templates, expected to be part of the conserved core, are averaged.

4.11
Building Structurally Divergent Regions

Step 4:
Model the Regions Outside the Conserved Core

The construction of the structurally divergent regions, called SDR, is a serious and open problem in comparative modeling, not only for large regions that have undergone local refolding during evolution, but also for relatively short regions where insertions and deletions occur. It is customary to call these locally refolded regions "loops", even if they do not necessarily correspond to loops in the protein structure.

First, very careful inspection of the alignment, and of the template structure, is a very important part of the procedure, because the first thing that must be done is to correctly position the gaps and one often has to modify the alignment manually and shift the gaps by a few positions to place them outside regions of secondary structure or of the packed core.

To build the SDR, we can rely on their sequence pattern, on the structure of the corresponding regions in other homologous structures, on the (albeit approximate) knowledge of the regions surrounding them that have been built by comparative modeling techniques, or on energetic calculations. We will discuss these approaches, although it is known that none is currently satisfactory. The problem is not only that the results of these procedures are often incorrect but, more importantly, that it is difficult, if not impossible, to have an a priori estimate of their accuracy. In other words, there are instances when they work and instances when they do not, but it is almost always impossible to tell which beforehand.

Local refolding usually occurs at the periphery of protein structures, in regions that are not subject to a very stringent evolutionary pressure and are, therefore, rarely involved in function, with some notable exception that we will discuss separately. This is good news, because even if we fail to model these regions correctly, it is likely that this will affect to a limited extent only our ability to interpret the information given by a model in terms of its biological significance. It is, nevertheless, a rather frustrating aspect of protein-structure prediction, and probably that which is most embarrassing to modelers.

If a loop is short (three-four residues) we can take advantage of the observed sequence pattern, compare it with the data reported in Table 1.1, and try to model the dihedral angles of its amino acids. As already mentioned, especially if the loop connects two antiparallel beta strands, i.e. if it is a hairpin, there is a correlation between the position of glycine amino acids and the conformation of the loop. This is not as useful as may seem at first sight. First, it is rare that we find insertions and deletions in these tight loops – their structural requirements are such that they are often conserved and rarely involved in local refolding. The other aspect is that, sometimes, tertiary interactions can overcome the sequence requirements and we will see an example of this phenomenon in the section dedicated to immunoglobulin loops.

When loops are medium-sized, it is much more difficult to define them according to their main chain dihedral angles, and consequently the classification becomes less rigorous. It is, however, possible to derive some approximate rules for the structure of these loops, on the basis of the types of interaction that stabilize them. For loops that form compact substructures the main factor determining conformation is the formation of hydrogen-bonds to main chain atoms of the loop. For loops with more extended conformation the required stabilization is obtained by packing an inward pointing hydrophobic side-chain of the loop between the secondary structure elements connected by the loop. The interesting question that arises is, of course, how conserved are such interactions and whether the stabiliz-

Figure 4.16 The figure shows two loops with similar conformations stabilized by the packing of a central hydrophobic amino acid. Note that one of the loops connects two alpha helices and the other two beta strands.

ing elements can be detected and used to predict the structure of the loop. Figure 4.16 shows one of the many examples that can be used to convince the reader that this is not so.

The structures of two very similar medium sized loops are shown in Figure 4.16. The sequences of the two loops are not similar. In both examples the loop is stabilized by a large and hydrophobic central side-chain which is packed into a cavity of the protein. The first loop connects two strands, however, and the second connects two helices, so it would be impossible to infer their structural similarity from either their local sequence or their context. The other example is shown in Figure 4.17 where there are three very similar loops without any clear local sequence similarity. Two of the three loops have a *cis* proline in an equivalent position and all are stabilized by hydrogen-bonds; such hydrogen-bonds are formed by the residues of the loop with completely unrelated partners, however. In the first loop the partner for these interactions is the side-chain of the residue preceding the loop; in the second it is the main chain of an alanine distant in the primary structure, in the third the propionyl group of a heme. In all these examples the hydrogen-bond partners occupy the same position in space relative to the loop. The structural context of these three loops is once again completely different. In the first two examples the loop is a hairpin, in the third it connects strands from different sheets.

The conclusion that can be derived from these examples is that the conformation of the loop dictates the interactions required to stabilize it, but in different proteins a variety of different topologies can be used to provide these interactions. This implies that it is unlikely that rules relating sequence to structure can be identified in medium sized loops.

When more than one homologous structure is available, the sequence and length of the region that we are modeling might be more similar to the corresponding region of a protein other than the best template and it is advisable in these instances to use the alternate structure as a local template.

As we have already mentioned, we have available the model of the regions surrounding a loop, usually called stems, and we can try to make use of the knowledge of their structure to model the intervening region. This approach is rather dated, but it still used and sometimes even successful. It originated from a method developed by Alwyn Jones from Uppsala University aimed at building fragments of proteins into electron density maps obtained by X-ray crystallography

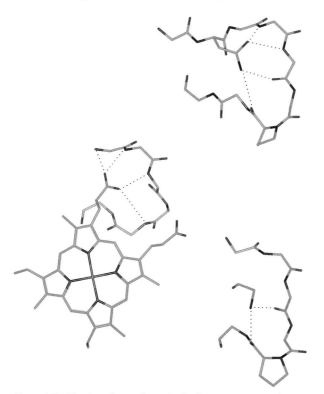

Figure 4.17 The three loops shown in the figure are very similar and stabilized by hydrogen-bonds, however the partners of these interactions are different in the three different proteins (an immunoglobulin, a viral protein, and a cytochrome).

experiments. In the original application of the method one searches in the database of proteins of known structure for regions that fit reasonably well the local electron density in the X-ray map. Clearly, the approach is more useful for loop regions, because regular elements of secondary structure are easy to build manually. The idea that stemmed from this approach is: given the amino and carboxy-terminal ends of a loop, how many ways are there for the amino acid chain to bridge between them? If their number is limited, we might hope that the correct one has already been observed in a known protein structure and, therefore, we can look for regions of known structure that are similar to the stems of the loop to be modeled and contain the correct number of amino acids. If, as almost always happens, more than one is found, we can use the sequence pattern or some energetic consideration to select the most likely.

For this method to work, two conditions should be verified – a loop similar to that to be modeled should be present in one of the proteins of the database of known structure and, if two loops are similar, their stems should also be similar, so we can use the structure of the latter to identify the former. While the first condition is very often verified, the second is not, as shown, for example, by the example depicted in Figure 4.16. Sometimes similar loops have similar stems, especially in evolutionarily related proteins, but more often they do not.

Another approach, that can be used for relatively short loops is not to rely on the database of known structure but to actually build, with a clever algorithm, all possible stereochemically reasonable loops that can bridge between the two stems. With either approach, a major problem is selection of which of the many loops is the most appropriate. Some of the hits can usually be discarded on the basis of their incompatibility with the rest of the structure (for example their main-chain atoms overlap main-chain atoms of the rest of the structure), but discriminating between the usually high number of other plausible alternatives turns out to be very difficult. Ideally, one should be able to compute the energy of the resulting protein and select the model with the lowest energy. We have discussed the limitations of energy-based evaluation of protein conformations earlier in this book, and it should not come as a surprise that this is not a very effective means of solving the problem.

4.12
A Special Case: Immunoglobulins

Predicting the structure of loops is difficult mainly because they are not subjected to a strong evolutionary pressure and therefore we are faced with an enormous number of possibilities with very few hints about what we are seeking. That this is so is proven by immunoglobulins, in which loops play a very important functional role and, indeed, can be predicted with respectable accuracy.

Immunoglobulins are multi-domain proteins consisting of two identical copies of a light, (L) and a heavy (H) chain, each including a variable domain. The antigen binding site is formed by six loops (denoted L1, L2, L3, H1, H2, and H3), clustering in space to form the antigen binding site as shown in Figure 4.18. The high

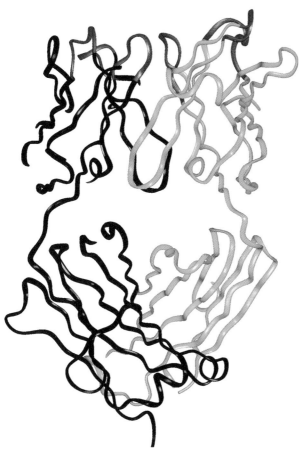

Figure 4.18 The structure of a fragment of an immunoglobulin. The antigen binding loops are shown in red.

sequence variability of these loops enables immunoglobulins to recognize a variety of different antigens. Comparative analysis of immunoglobulin structures has revealed that different sequences in different antibody loops do not always generate different conformations in both the main chain and side-chains of these regions. Five of these six loops can only assume a limited number of main-chain conformations, called "canonical structures". Most sequence variations only affect the side-chains of the loops, consequently modifying the antigen binding-site surface, without changing the backbone structure of the loop. Only some specific sequence changes, in a limited set of positions, produce a change in the main chain conformation of the loops.

Careful analysis of the available structures highlighted the special relationship between sequence changes and canonical structures. There are specific residues which, as a result of their packing, hydrogen-bonding, or ability to assume unusual values of their main chain dihedral angles, are responsible for the occurrence of each canonical structure. This implies that it is possible, for five of the six immunoglobulin hypervariable loops, to define precisely the sequence-structure relationship. As an example, Figure 4.19 shows the canonical structures for six-residue L3 loops of immunoglobulins. Important determinants for the occurrence of one or the other conformation are the position of a proline residue and the nature of the interaction of the residue preceding the loop with atoms of the loop itself. In the canonical structure on the left in the figure, there is a proline in position 95 (according to the common immunoglobulin numbering scheme devised by Elvin A. Kabat and named after him) with a *cis* peptide bond (recall that this means that the dihedral angle Ω around the peptide bond is approximately $0°$) and the side-chain of the residue immediately preceding the loop forms hydrogen-bonds with the main chain atoms of the loop. In the other canonical structure the proline is, instead, in position 94 and its peptide bond is in a *trans* conformation (Ω close to $180°$). The position of the proline is, therefore, diagnostic of the canonical structure of the loop and it can be effectively used to predict its conformation.

Analogous situations are observed for all the other loops except for the central region of H3, for which such a clear sequence structure correlation has not been found. These loops are certainly a special case and their structural features are quite unique to this class of molecule.

The second loop of the heavy chain (H2) is a beta hairpin, often short (3 or 4 residues). The expectation, from the discussion in Chapter 1 about hairpin loops, is

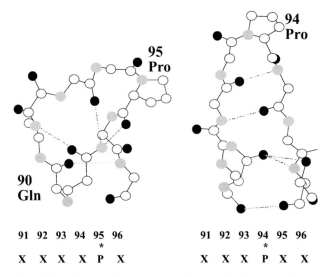

Figure 4.19 The canonical structures of immunoglobulins. The loop shown in the figure is called L3 (it is the third loop of the light (L) chain of antibodies and is part of the antigen-binding site). When the length of the loop is six amino acids, as in the figure, only two main chin conformations are observed. The one on the left occurs when the amino acid in position 95 is a proline. The conformation shown on the right instead occurs when the proline is in position 94. All other residues are free to vary and contribute to the shape of the antigen binding region.

4.12 A Special Case: Immunoglobulins

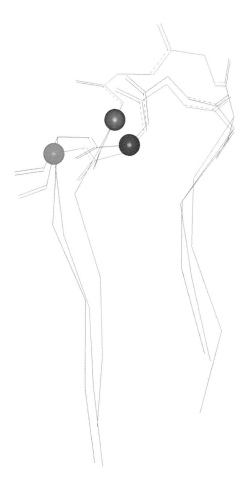

Figure 4.20 Superposition of the H2 hairpin loops of three immunoglobulins. Their conformation does not follow the rules relating sequence and structure in hairpin loops. The determinant of their conformation is the type of amino acid that occupies position 71, and not the position of the glycine (indicated by the sphere in the figure).

that for these short loops there is a correlation between their sequence pattern and their conformation.

An interesting result was obtained by comparing H2 loops of different immunoglobulins of known structure, for example those of 2FBJ, 1NCD, and 2FB4.

All these form a four-residue hairpin turn; their sequence is shown in Table 4.1. If one observes the positions of the glycine units and recalls the discussion in Chapter 1, the conclusion should be that the conformation of the loop of 1NCD and that of 2FBJ should be very similar in that both have a glycine in the fourth position, and different from that of 2FB4, where the glycine is in the second position. What is instead observed is that the conformation of the 2FBJ loop is much more similar to that of the corresponding 2FB4 loop. Careful analysis of the interactions of this loop with the rest of the immunoglobulin structure shows that the determinants of the conformation of this loop involve tertiary interactions, in particular they depend on the size of residue 71, a residue far away in the sequence from the loop, and part of the conserved immunoglobulin conserved framework. When position 71 contains a small or medium sized residue, for example leucine, as in 1NCD, the conformation of four-residue H2 loops is similar to that illustrated in blue in Figure 4.20; when residue 71 is arginine, as in 2FBJ and 2FB4, different packing of the side-chains arises in the loop region and the main chain of the loop has the conformation illustrated in red and green in Figure 4.20. The implications of this observation are many and relevant both to our understanding of loop architecture and for practical purposes. First, as already stated, a tertiary interaction can be responsible for loop conformation, so our ability to predict the structural conformation of short hairpins may very well be impaired by our limited understanding of the overall stability requirements of proteins. Second, the ability to transplant loop regions from one protein to another, for example from antibodies of non-human origin into human frameworks, relies on the assumption that the conformation of the loops is independent of the rest of the structure, which is obviously not generally true.

Table 4.1 Sequence of the H2 loop of three antibodies.

1NCD	T	N	T	G
2FBJ	P	D	S	G
2FB4	D	G	S	D

4.13
Side-chains

Step 5:
Model the Position of the Side-chains of the Target

Side-chains interact with each other and the energy contribution of their interactions is an important aspect of the stabilization of the native conformation of a protein. The problem of finding the correct combination of dihedral angles of the side-chains, however, is combinatorial in nature. We should inspect every possible combination of side-chain dihedral angles and select the optimum one by optimizing a target function that represents their interaction energy. The energy of the conformation of a side-chain depends upon its tertiary interactions with the rest of the protein structure. However, as originally noted in 1987 by Ponder and Richards, some amino acids have preferences for specific side-chain conformations. The frequencies of the preferred conformations for each amino acid are reported in tables called rotamer libraries. Backbone-independent libraries report the frequencies of each amino acid side-chain conformation, irrespective of the values of their main chain angles. In backbone-dependent libraries the rotamer frequencies are computed as a function of the main chain dihedral angles of the amino acids. Most methods for predicting the side-chain conformations use the conformation with highest frequency for each amino acid as initial values and subsequently modify them to optimize the energy of their combination. Because conserved side-chains tend to retain their conformation in homologous proteins, their side-chain angles are usually taken from the template, rather than from the rotamer libraries.

The search for the optimum conformation of the side-chains in the model requires a search strategy and a target function to be optimized. The target function is usually an energy function, sometimes simply taking into account steric interactions or using knowledge-based potentials. More specific to the problem of side-chain prediction is the search strategy, which needs to be fast and efficient. The search method can be exact or approximate. Only when it is exact is there a guarantee that the optimum of the target function, for example the knowledge-based potential energy, is found. Approximate methods do not necessarily find the global optimum, but they can be clever enough to reach conformations reasonably close to the one with minimum energy.

The algorithm most used for placing side-chains in a model, and one of the most efficient, is SCWRL. In short, this method first positions each side-chain in its most favorable conformer, unless this is sterically incompatible with part of the modeled main chain other than that of the amino acid itself. If this is so, the conformer is discarded and the next most frequent conformation for the amino acid is chosen, iteratively. If the atoms of some side-chains are too close to other side-chain atoms, all the amino acids involved in the unfavorable interactions are labeled as "active residues" and clustered. This means we group the residues with unfavorable interactions among themselves, but none with amino acids outside the cluster. Their conformation can now be locally optimized by the search procedure that, for SCWRL and most side-chain-modeling procedures, is based on a dead end elimination strategy.

Briefly, the dead end elimination strategy, which is an optimization procedure that can be also used for applications other than side-chain modeling, is based on the idea that there are some values of the rotamers that are incompatible with the global energy minimum conformation. We can skip, in the search, rotamers of one residue if another rotamer for the same residue always has a lower-energy interaction energy with all other side-chain and main-chain atoms of the protein, irrespective of which rotamer is chosen for the other side-chains.

4.14
Model Optimization

The steps outlined above – template selection, construction of the main chain of the core, prediction of the loop regions, and positioning of the side-chain – provide us with an approximate model of the target protein. Ideally we would now like to optimize the resulting structure on the basis of energy calculations. In CASP experiments, it is possible to submit pairs of models for the same target, one representing the structure before optimization and the other representing the structure after optimization. The goal is to verify whether optimization procedures are able to modify the starting model to make it more similar to the experimental structure of the target protein. So far the results have been rather unsatisfactory and no method seems to be able to consistently improve over a starting model.

4.15
Other Approaches

The step-wise procedure outlined above is, as already mentioned, not necessarily ideal for achieving a sensible model of our target protein. There are two approaches that do not strictly follow this classical strategy. The very popular method Modeller constructs the complete models on the basis of spatial constraints. In other words, it computes a set of distance and dihedral angle probability distributions that must be satisfied by the final models and then builds the models that are compatible with these distributions. The probability distributions are derived from a detailed analysis of family of proteins of known structure. For example, one can compute the probability of observing a certain $C\alpha-C\alpha$ distance in a protein, given the observed distance in a homologous protein, the type of amino acids, the dihedral angles, the sequence similarity between the two proteins, etc.

The spatial restraints and some energy terms to ensure proper stereochemistry are combined into a target function that is optimized by simulated annealing. Several different models can be calculated by varying the initial structure. The local variability among these models can be used to estimate the errors in the corresponding regions of the model.

The most recent developments in comparative modeling are based on the idea of constructing several models for each target protein and selecting the most likely only at the end of the complete model-building procedure. In other words, rather than optimizing independently each of the steps of the procedure, the most successful methods funnel into each subsequent step not only the optimum but also the suboptimum intermediate results. Selection of the final model is based on the analysis of the several resulting atomic structures. To some extent this is as if all the stages of the model building procedure (template selection, alignment quality, local templates for insertions and deletions, and side-chain positioning) were optimized at the same time rather than sequentially. The strategies for building several alternative models include selection of templates not only on the basis of sequence similarity, but also on the basis of sequence-structure fitness evaluation, taking advantage of algorithms developed for fold-recognition prediction methods. Sometimes, the template originally selected is used for searching the data base of structurally related proteins to select folds that can be used as alternative templates. Both optimum and suboptimum sequence alignments with each of the putative templates are used as the basis for model building and additional three-dimensional models are sometimes generated by combining fragments of the obtained models.

Evaluation of the final set of models, after a structure clustering step, can be based on several independent criteria, for example evaluation of local environment and inter-residue contacts, knowledge-based pairwise potentials, stereochemical quality, and, occasionally, visual inspection.

These methods can produce, on average, better models than those obtained by conventional step-wise modeling procedure, probably because it is more effective to evaluate the quality of a final three-dimensional complete model than that of each of the intermediate results of the procedure.

4.16
Effectiveness of Comparative Modeling Methods

In the last chapter we will describe some successful practical applications of comparative modeling in which a combination of methods, expertise, experimental data, and careful analysis of the model has led to significant progress in the understanding of biological systems. Here we will review what we have learned from the various CASP experiments, in which the submitted models are usually produced in a very short time and therefore reflect more faithfully the state of the art in the field.

The overall picture is that models with very respectable accuracy can be built for proteins sharing a significant similarity with proteins of known structure. A sequence identity between target and template of more than 40% essentially guarantees that the overall structure is correctly predicted by most, if not all, techniques (Figure 4.21).

When a more distant relationship exists between the protein under study and the closest protein of known structures, the gap between different methods becomes

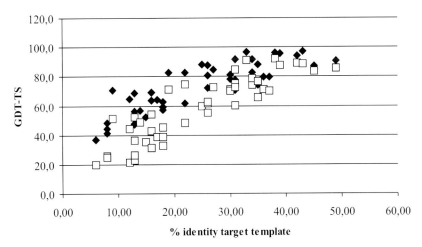

Figure 4.21 The relationship between the GDT-TS of the best (filled symbols) and average (open symbols) models and the sequence identity between the target protein sequence and the sequence of the best structural template. The data are taken from the CASP5 results and indicate that, above 40% sequence identity between target and template sequence, most methods can produce very respectable models. In more difficult examples the best methods can still produce useful results, but the gap between the quality of their results and those that can be obtained on average increases.

more apparent. As we mentioned, techniques that build several models and, only at the end of the procedure select the final method, perform better. This is not only true for "human" predictions, i.e. for predictions submitted by research groups, but also for predictions obtained by automatic servers. Some of these, called meta-predictors, use a similar strategy. Rather than building a model, these servers outsource the prediction to several other servers, collect the results and either combine or score them to provide the user with a final model. Although these methods work, on average, better than single servers, one should not forget that they can exist only insofar as "regular" servers keep being developed and improved.

Prediction of the structure of loops, especially if longer than a few residues, is still an open problem and accuracy is not yet satisfactory. What is worse, although it is possible to estimate, a priori, the accuracy of the prediction of the backbone of a protein on the basis of the sequence similarity between the target and template(s) proteins, there is no clear way to learn beforehand whether or not a given loop will be correctly predicted by any method; this is a serious limitation. The accuracy of the prediction of side-chains is rather difficult to estimate. The number of side-chains that one can include in the evaluation is quite limited. They should be those in targets solved with high accuracy by X-ray crystallography, not exposed to the solvent, with low B factors, and not involved in crystal contacts. One accepted conclusion is that the accuracy of side-chains is very dependent on the accuracy of backbone prediction. In other words, the better the prediction of the backbone, the better methods for building side-chains work. This implies that improvement of the quality of the prediction of the backbone will produce, by itself, an improvement in the prediction of the side-chains.

Finally, as already mentioned, there is no consistent example of a method able to improve a model with energy-based methods, such as energy minimization or molecular dynamics. Anecdotal examples are present in the literature and have been observed for some CASP predictions, but it is still true that these methods rarely succeed in modifying a model in such a way as to make it closer to the experimental structure.

CASP results clearly demonstrate without doubt that there has been progress in the field; we should not, however, forget that, between one experiment and the next, the sequence and structure databases continue to grow and this, by itself, makes predictions easier. A larger structure database not only increases the likelihood of finding a good template for the target protein, but also enables better understanding of the structural variability of the template family and therefore a better definition and prediction of the core of the protein. More importantly, model-building procedures rarely make use of two sequences alone. Usually the alignment used for model building is extracted from a multiple sequence alignment of proteins of the same family, and its quality will depend upon the number and the similarity distribution of all the sequences in a multiple sequence alignment. A larger sequence database enables more sequences of the target and template family to be included in the multiple sequence alignment and this is likely to improve the alignment, as we discussed. Indeed, this latter effect can be taken into account at least partially when comparing results between different editions of the CASP experiment.

For example, one can analyze the multiple sequence alignment available at the time of each experiment for each target, calculate the pair-wise sequence identity between each pair of sequences and use the values to construct a graph similar to that shown in Figure 4.22, in which each node represents one of the sequences in the multiple sequence alignment and the lengths of the edges are proportional to the distance (inversely proportional to the percentage of identity) between the connected nodes. The multiple sequence alignment is a path in the graph that includes all the sequences.

The difficulty of aligning target and template depends upon the availability of intermediate sequences, and this is determined by the most difficult pair-wise alignment that we need to perform to go from the target to the template. In other words, we might end up aligning a target and a template sequence only sharing a very low sequence identity, but we might achieve this by aligning pairs of very similar intermediate sequences, starting from the target and "jumping" from one sequence to the other until we reach the target, much in the same way as we might cross a large river jumping from one stone to the next. The difficulty of crossing the river is not proportional to its width but to the longest jump that we need to make between two stones.

Given all possible paths including target and template, we are therefore interested in those in which the maximum distance between each pairs of traversed nodes is minimal (Figure 4.22). When such a path is found, the longest edge in the path, i.e. the sequence similarity between the two most diverse sequences in the path, is an estimate of the difficulty of aligning target and template, given the distribution of sequences in the multiple sequence alignment. This approach gives,

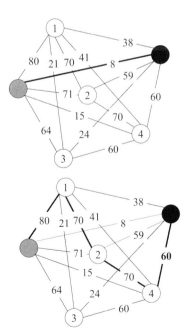

Figure 4.22 Graphical scheme of a method for evaluating the difficulty of aligning two protein sequences when a multiple sequence alignment is available. In the scheme, each circle represents a protein and each edge is labeled with the sequence identity between the two connected proteins. Assume that the gray circle is the target protein and the black circle the template. The sequence identity between the two protein sequences is only 8%. We can, however, progressively align the proteins following the path indicated by the ticker lines in the lower part of the figure. In this instance the most difficult alignment that we are forced to perform is that between the protein labeled "4" and the template sequence.

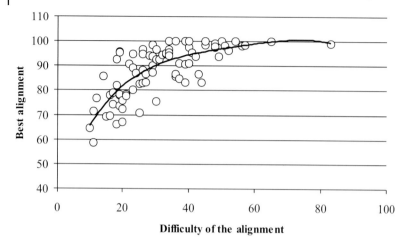

Figure 4.23 Relationship between the difficulty of aligning a target and template protein sequences, computed as described in the legend to Figure 4.22, and the best alignment obtained in the CASP experiments for the same pair of sequences.

to a first approximation, a measure of the difficulty of aligning the target and template sequence for each target in different experiments, given the database available at the time of the prediction, and can be used to ask whether the alignment of targets and templates of equivalent difficulty has become more accurate with time. The answer is that any improvement in alignment among different CASP experiments is mainly because of the availability of more sequences, and not because of a genuine improvement of the methods (Figure 4.23). In other words, targets of equivalent difficulty, are aligned with similar accuracy in all CASP experiments (in fact, in all CASP experiments subsequent to the introduction of effective methods for multiple sequence alignments).

Another frustrating aspect that arises from analysis of CASP experiments is that no method yet seems very effective at correctly modeling multi domain proteins when the domains are modeled using different templates. In other words, each domain can be correctly modeled taking advantage of its similarity with domains of proteins of known structure, but it is not easy to predict their relative orientation in the target protein.

In all the cases analyzed, the region of the active site of enzymes is, on average, better predicted than the rest of the protein structure. It is generally believed that this reflects more the intrinsic higher conservation of these regions than specific aspects of the methods. By its own nature, comparative modeling exploits the evolutionary constraints posed by the biological function upon a protein and is, therefore, expected to work better on regions that are well conserved. Nevertheless, it is still important to bear in mind that functionally important regions are likely to be well predicted by comparative modeling. This is, in fact, the reason why this method, with all its pitfalls and problems, is an invaluable tool in modern biology (Figure 4.24).

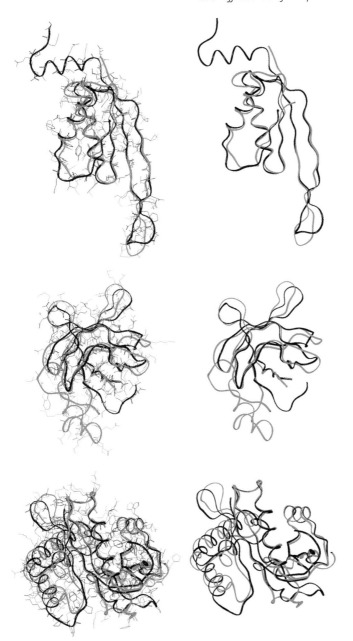

Figure 4.24 Some examples of predictions obtained by comparative modeling techniques in the CASP experiments. The experimental structures are shown in blue and the models in green in all three examples. On the left both structures are shown with their side-chains. The percentages of identity between the cores of the target protein and the best available template are 19%, 27%, and 10%, respectively. The difficulty, defined in Figure 4.22, is 26%, 27%, and 18%. Note that in all the examples the peripheral parts of the proteins are predicted less accurately.

Suggested Reading

In this chapter we mentioned several tools, most of them available via the Internet. We will give the reference to the original article. Most of the tools described here can be found at either http://ncbi.nlm.nih.gov or at http://www.ebi.ac.uk. For others, whenever applicable, the location of the server where the tool can be found is listed after the reference.

The alignment algorithms:
S.B. Needleman, C. D. Wunsch (1970) A general method applicable to the search for similarities in the amino acid sequence of two proteins. J. Mol. Biol. 48, 442–453
T. Smith, M. Waterman (1981) Identification of common molecular subsequences. J. Mol. Biol. 147, 195–197

The substitution matrices:
M.O. Dayhoff, R. M. Schwartz, B. C. Orcutt (1978). A model for evolutionary change. In: M. O. Dayhoff (Ed.) Atlas of Protein Sequence and Structure, Vol. 5, National Biomedical Research Foundation, Washington, pp. 345–358
S. Henikoff, J. G. Henikoff (1992) Amino acid substitution matrices from protein blocks. Proc. Natl. Acad. Sci. USA 89, 10915–10919

Database search programs:
Altschul, S. F., Gish, W., Miller, W., Myers, E. W. and Lipman, D. J. (1990) Basic local alignment search tool. J. Mol. Biol. 215, 403–410
Pearson W. R. and Lipman D. J. (1988) Improved tools for biological sequence comparison. Proc Natl Acad Sci USA 85, 2444–2448
Altschul, S. F. and Koonin, E. V. (1988) Iterated profile searches with PSI-BLAST– a tool for discovery in protein databases. Trends Biochem. Sci. 23, 444–447

Alignment methods:
D.G. Higgins, J. D. Thompson, T. J. Gibson (1996) Using CLUSTAL for multiple sequence alignments. Methods Enzymol. 266, 383–402
C. Notredame, D. G. Higgins, J. Heringa (2000) T-Coffee: a novel method for fast and accurate multiple sequence alignment. J. Mol. Biol. 302, 205–217

The PFAM collection of hidden Markov models:
A. Bateman, E. Birney, R. Durbin, S. R. Eddy, K. L. Howe, E. L. Sonnhammer (2000) The Pfam protein families database. Nucleic Acids Res. 28, 263–266; http://www.sanger.ac.uk/Software/Pfam/

The original idea of using fragments of known protein structures:
T.A. Jones, S. Thirup (1986) Using known substructures in protein model building and crystallography. EMBO J 5, 819–822

The original paper tabulating the frequency of side-chain conformation:
J. Ponder, F. Richards (1987) Tertiary templates for proteins. Use of packing criteria in the enumeration of allowed sequences for different structural classes. J. Mol. Biol. 193, 775–791

The tools for building side-chains:
R. Dunbrack Jr, M. Karplus (1993) Backbone-dependent rotamer library for proteins. Application to side-chain prediction. J. Mol. Biol. 230, 543–574; http://dunbrack.fccc.edu/SCWRL3.php
L. Holm, C. Sander (1992) Fast and simple Monte Carlo algorithm for side-chain optimization in proteins: application to model building by homology. Proteins: Structure, Function Genetics 14, 213–223; http://www.cmbi.kun.nl/gv/hssp/

The prediction of loops:
C. Wilmot, J. Thornton (1988) Analysis and prediction of the different types of beta-turn in proteins., J. Mol. Biol. 203, 221–232
A. Tramontano, C. Chothia, A. M. Lesk (1989) Structural determinants of the conformations of medium-sized loops in proteins. Proteins: Structure, Function and Genetics 6, 382–394

Immunoglobulins and their loops:
C. Chothia, A. M. Lesk, A. Tramontano, M. Levitt, S. J. Smith Gill, G. Air, S. Sheriff, E. A. Padlan, D. Davies, Tulip W. R. (1989)

Conformations of immunoglobulin hypervariable regions. Nature **342**, 877–883

A. Tramontano, C. Chothia, A. Lesk (**1990**) Framework residue 71 is a major determinant of the position and conformation of the second hypervariable region in the VH domains of immunoglobulins. J. Mol. Biol. **215**, 175–182

Progress in comparative modeling methods is described in one paper in each of the CASP issues (http://predictioncenter.llnl.gov) and in:

D. Cozzetto, A. Tramontano (**2005**) The relationship between multiple sequence alignment and the quality of protein comparative models. Proteins **58**, 151–157

5
Sequence-Structure Fitness Identification: Fold-recognition Methods

5.1
The Theoretical Basis of Fold-recognition

Although the number of known protein structures and sequences grows at an impressive rate, still too often we face the problem of having to infer structural properties of proteins for which no homologous protein of known structure is available. In such circumstances we cannot use comparative modeling methods and are left with a sequence, or a family of sequences, for which no structural information is available. Methods that can be used to try and infer some structural properties, for example secondary structure or pairs of amino acids putatively in contact from a multiple sequence alignment, will be discussed in chapter 7.

If, however, we analyze the dataset of available structures, we can derive another property of protein structure that can be used to produce models of unknown proteins – the folding code is degenerate in that proteins that do not seem to share an evolutionary relationship can have a similar structure.

Let us assume we compare with each other all the sequences of proteins of known structure, group them into evolutionary families and select only one protein per family. If the relationship between sequence and structure were a one-to-one relationship, each of the selected proteins would have a different architecture, but this is not observed. The relationship is many-to-one in the sense that many seemingly unrelated proteins share a similar fold (Figure 5.1).

> *Question: Are all folds equivalently used by nature?*
>
> »The distribution of folds is highly non uniform. A handful, approximately ten, are shared by a large number, about 30%, of known proteins with large diversity in sequences and functions. We call these "superfolds". Among these we find the immunoglobulin fold, shared by many domains of membrane receptors, the Rossman fold, the TIM barrel, and the hemoglobin arrangement. Although it is relatively easy to

Protein Structure Prediction. Edited by Anna Tramontano
Copyright © 2006 WILEY-VCH Verlag GmbH & Co. KGaA, Weinheim
ISBN: 3-527-31167-X

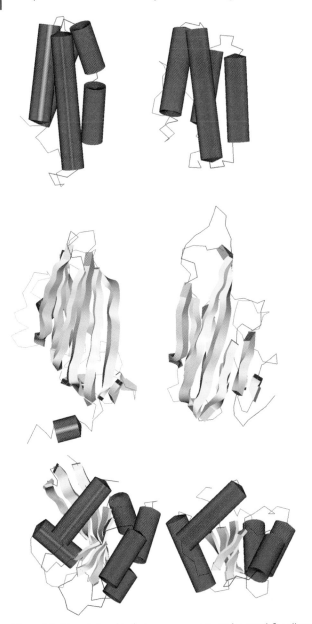

Figure 5.1 The relationship between sequence and structure is degenerate. Three pairs of apparently unrelated proteins having a similar architecture are shown in the figure. The pairs (top to bottom) are: hemerythrin (an oxygen-transporting protein) and a cytochrome B_{562} (involved in electron transport); ras p21 (an oncogene) and CheY (a protein involved in bacterial flagellum motion); a protein of the satellite tobacco necrosis virus and a tumor necrosis factor. Note that the overall topology of the proteins of each pair is similar but the size of the elements of secondary structure may differ and some peripheral extra elements can be present in one protein but not in the other.

identify close homologous relationships, as discussed in
Chapter 4, it is more complex to define precisely what we
mean by a "common fold", therefore establishing exactly how
many "superfolds" are present is not straightforward. The fact
remains that nature seems to use some structural arrangements more often than others. This might be because of the
inherent thermodynamic stability of the fold and/or to the
prevalence of common recurring structural motifs.«

How similar are the structures in this case is rather difficult to answer, because we have nothing with which to correlate the similarity; certainly the structures will be more different than among evolutionarily related proteins (Figure 5.1), although often one of them can still represent a useful starting point for modeling the analogous one and this is the basis of the "fold-recognition" methods we will describe in this chapter. These methods attempt to detect which, if any, of the known folds can be adopted by a target protein. Although there are many approaches to the problem, the unifying theme is that they try and find folds that are compatible with the target sequence. This is a different way of formulating the prediction problem – rather than asking what is the structure of a target protein, these methods ask whether any of the known structures can represent a reasonable model for it, irrespective of the existence or detectability of an evolutionary relationship. The problem is how to evaluate the fitness function of a sequence and a structure and, usually, this can be done using two alternative approaches: "profile based methods" and "sequence threading".

5.2
Profile-based Methods for Fold-recognition

The basic idea of profile-based methods for fold recognition is that the physicochemical properties of the amino acids of the target protein must "fit" with the environment in which they are placed in the modeled structure. For example, if a known structure is used as a template to model the target sequence, are hydrophobic residues located inside the protein? And are residues likely to be found in beta strands, for example beta-branched residues, indeed placed in a beta strand? There are complications, however. A charged residue placed in the core of the protein is clearly in an unfavorable situation, unless another residue of opposite charge is nearby and can form a salt bridge with it. In other words, we should not only evaluate the likelihood that a given residue is favored in a protein position, but also take into account the putative interactions that the residue can establish with other residues in the final structure. Although this is not taken into account in profile-based methods, they can often recognize the correct fold.

Profile-based methods code each amino acid of the target sequence according to its properties, for example secondary structure propensity, hydrophobicity, average accessibility in protein structures. Structure and accessibility propensity can be derived from statistical analysis of known protein structures, by counting how often each residue is found in helices and strands and how often it is exposed to solvent. In other implementations, they can be obtained, for the specific protein sequence, by use of secondary structure and accessibility prediction methods.

The next step is to analyze each known protein structure, or a properly selected subset of these, and assign to each position in the structure a symbol coding its environment (secondary structure, exposure to solvent, number of hydrophobic contacts with other residues) irrespective of the amino acid that happens to occupy that position in the specific protein structure. In this way, a protein structure is encoded as a string of symbols comparable with the symbols we have used to encode our target sequence. In the sequence each symbol reflects the propensity of the corresponding amino acid for a certain environment, and therefore it is related to the specific amino acid, while in the string encoding the structure, the symbol reflects the actual environment of each position, and the same type of amino acid can be represented by different symbols in different parts of the structure (Figure 5.2).

All that is needed now is to compare the two strings in much the same way we compare protein sequences. In this procedure we must again allow insertions and deletions and assign penalty values to them, define a score and compare the observed score with the expected background distribution of scores. Methods

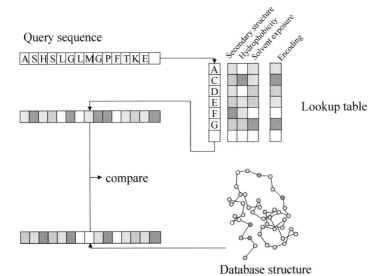

Figure 5.2 Schematic diagram of a possible profile-based method for fold recognition. The amino acids of the query sequence are replaced by a code that summarizes their hydrophobicity and their propensity for secondary structure type and solvent exposure. Each structure in the database is also encoded as a string by assigning a code to each of its amino acid positions. The code reflects their structural environment (secondary structure, solvent accessibility, and hydrophobicity of their environment). This does not depend on the actual amino acid present in the position analyzed. The string encoding the query sequence and each of the strings encoding the database structures are aligned and compared.

very similar to those used in sequence alignment are used here. The procedure can be made more sophisticated, for example by using a multiple sequence alignment of proteins of the family of the target protein and constructing a profile of propensities taking into account the variability of each position. Sometimes this procedure generates alignments between the target sequence and the "recognized" structure that point toward an evolutionary relationship, albeit distant, between the two proteins. The relationship might have been very difficult to detect when the protein sequence was compared with the whole database of known sequences, but it can become apparent when fold-recognition-based alignment points at conserved residues or regions. For this reason, in CASP it is customary to divide the targets into "homologous" and "analogous" fold-recognition targets. What the first term means is that, although it was very difficult or impossible to detect the evolutionary relationship between the target and the template on the basis of their sequences, subsequent comparison of their structures revealed common features that can be explained only by invoking an evolutionary relationship.

The distinction between comparative modeling targets and fold-recognition targets is, consequently, very fuzzy. Very sensitive methods for homology detection, for example hidden Markov models, can therefore be regarded as fold-recognition rather than alignment or database-searching techniques.

5.3
Threading Methods

The term "threading" was first coined in 1991 by David Jones, Janet Thornton, and Willie Taylor. Threading uses an atomic representation of protein structures and tries to "thread" a sequence of amino acid side chains on to a backbone structure (a fold) and to evaluate its fitness with the proposed template structure. The last step is usually based on knowledge-based pairpotentials (Chapter 3), that were indeed originally developed for this purpose, and on a solvation potential. The solvation potential is an important part of the energy function used in a threading method, because protein-solvent interactions play a major role in folding. Inclusion of solvent effects is also essential in calculations involving docking of proteins and ligand binding, and in protein recognition, engineering, and design. The precise calculation of the solvent contribution to energy is a problem, because the large number of degrees of freedom of water molecules makes the explicit simulation of water difficult. One means of estimating the solvation energy is to express it in terms of the reduction in the protein's solvent-accessible area of folding multiplied by the solvation free energy per unit area:

$$\Delta G_{\text{solv}} = \sum_i \sigma_i \Delta A_i$$

where σ_i is the atomic solvation parameter of atom i of a given type and ΔA_i is the change in solvent-accessible surface area upon folding. The atomic solvation parameters are atom-type parameters usually determined by least squares fitting of experimentally observed changes in free energies upon transfer of model compounds from vacuum (or a hydrophobic medium) to water. Several sets of atomic solvation parameters are available and it is still not established whether any is clearly superior to the others.

Given the energy function, we must use a search procedure that finds the optimum alignment between the query sequence and each structural template. In this process the optimum alignment is that which minimizes the energy of the new sequence in the template structure.

The pair interaction potential is non-local, so the energy corresponding to an alignment depends on all the interactions implied by the alignment (Figure 5.3). In sequence alignment the score for aligning two amino acids does not depend on the rest of the alignment, but only on the identity of the two amino acids. In threading, things are rather more complex, because we need to align the amino acids of the target to the amino acids of the template, but the score depends on the identity of the amino acids in neighboring positions which, in turn, depends on the final alignment.

Several techniques, for example double dynamic programming or Monte Carlo optimization can be used to address this issue, but the problem remains complex and computer-intensive. Some methods make use of the so-called frozen approximation in which the interaction partners are taken from the template protein rather than from the target (Figure 5.4). In other words, these methods place an

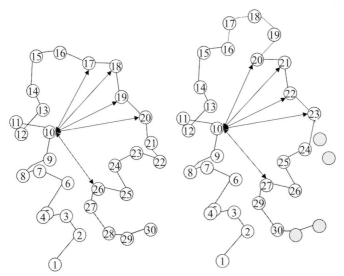

Figure 5.3 A query sequence can be positioned in a database structure in several ways, because there can be inserted and deleted residues, as shown in the right side of the figure. The interactions made by one amino acid, for example the one indicated with "10" in the figure, depend on the alignment of the rest of the sequence – the interactions of this amino acid (some of which are shown as arrows) are different in the two examples, reflecting two different alignments.

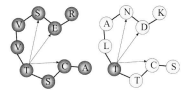

Figure 5.4 Schematic explanation of the frozen approximation. On the left, a database structure is shown with its original sequence (indicated by dark circles). In the right, the query sequence is positioned in the database structure in one of the many possible alignments. Calculation of the score should take into account which residues of the target sequence are in contact with, say, the threonine in the final alignment. In the frozen approximation, the interactions are computed by leaving the original sequence in every position of the database structure, except for the position occupied by the threonine. The procedure is repeated by substituting, in turn, each amino acid of the query sequence into a position of the target structure.

amino acid of the target sequence in a given position of the template structure and compute the energy by adding its interaction energy with the amino acids of the template. The underlying hypothesis is that, when an amino acid is placed in a given position, its interactions with the rest of the structure are approximated by its interactions with the amino acids of the template. The fitness is computed without taking into account the fact that the target amino acids will replace the template amino acids in the final model. This approximation is justified by the assumption that, if the target and template protein structures are similar, it is likely that many interactions will be conserved among them.

The procedure must be repeated for each of the templates in the selected set of folds and the resulting sequence-to-structure alignments must be sorted to decide which, if any, of the available structures is an appropriate template for the target. The results are often expressed in terms of Z-score, i.e. the number of standard deviations above the mean computed over the whole set of threading results.

If the target protein belongs to a family, it is a good idea to repeat the threading procedure using as input the sequences of other members of the family, which are expected to share the same fold. We wish to assess the compatibility of the target sequence and the template structure independently of the presence of a sequence similarity between them. It is therefore advisable to use, as input, sequences of the family as dissimilar as possible from the original target. If the same fold is found using two very distantly related sequences, our confidence in the result increases. In general, one expects to find more than one structure belonging to the same fold class in the high ranking positions. Another good indicator of correct detection of the fold seems to be a clear score separation between this first set of hits and subsequent hits.

A threading output consists of one structure, the selected template, two sequences, that of the target and that of the template, and their alignment. Careful analysis of the alignment can be used to increase the confidence in the result because, here also, there is still the possibility that a very distant evolutionary relationship between the two proteins can be detected at this stage and this makes our case for selection of the template much stronger.

5.4
Profile–Profile Methods

To improve the detection of related proteins it is often useful to include evolutionary information for both the target and template proteins, for example constructing profiles for both families and subsequently comparing the profiles. A typical scheme of a profile–profile method is to run a database search with a method such as PSI-BLAST. Usually several iterations of the program are run with an E-value threshold between 10^{-3} and 10^{-1}, depending on the method. Next, the significantly similar sequences are collected. Different algorithms differ in how they assign weights to each sequence and how they treat positions with a low level of sequence variation, deemed to be less informative. The simplest solution is to average all sequences from the multiple alignment with equal weights. Others perform a filtering procedure removing highly identical sequences, leaving only a set for which the similarity of all sequences is below a threshold (usually between 95 and 98%). The profile is then compared with pre-computed profiles including sequences of homologous proteins of known structure.

The methods essentially differ in the strategy used to assign scores to the comparison. Some methods use the sum-of-pairs score that we have already mentioned. A profile can be seen a set of vectors, one for each position of the alignment, and the two sets of vectors corresponding to the query profile and the precomputed target profiles can also be compared by computing their dot product or correlation coefficient.

5.5
Construction and Optimization of the Model

The final step of a fold-recognition procedure is actual construction of a three dimensional model of the target protein on the basis of the selected template and of the alignment provided by the procedure. The alignment can be optimized taking into account other factors. For example, one can predict the secondary structure of the unknown protein and compare it with the known secondary structure of the template. As we will see, the accuracy of methods for predicting secondary structure elements in proteins is quite high and the match between the secondary structure observed for the template and that predicted for the target can be used to evaluate the likelihood of the match. A word of caution is needed here, however – some threading methods use results from secondary structure prediction method to filter from the library of folds structures that are unlikely to be correct. In these circumstances, matching of secondary structure is a much less stringent criterion for the quality of the results, because it has already been used to pre-select the output.

Until a few years ago, fold recognition methods often detected the correct fold but gave a rather incorrect alignment. Because the fit of the sequence to the fold is evaluated on the basis of the alignment, this observation was rather puzzling. One

Figure 5.5 Some examples of predictions obtained by fold-recognition procedures in the CASP experiments. The experimental structures are shown in blue, the models in green. The first two proteins are examples of homologous fold recognition, the last of analogous fold recognition.

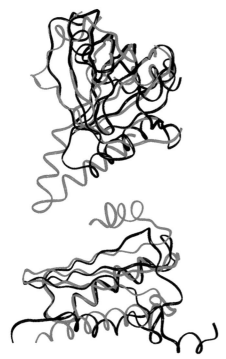

possible explanation is that early fold-recognition methods were able to recognize some general properties of a fold, such as number of hydrophobic residues, sequence length, etc., rather than the specific sequence-structure combination. Matters have changed quite substantially in more recent times. By combining sequence-based methods and structure-fitness tools, current methods are rather good not only at recognizing the fold but also in producing a reasonably good sequence alignment.

When a satisfactory alignment has been obtained, the steps to follow to obtain a complete set of coordinates are the same as those employed in comparative modeling. It should be kept in mind, however, that the expectation is that the final model will, usually, be of lower quality than those that can be obtained by comparative modeling and this should be taken into account when using tools to evaluate the correctness of the model.

Fold-recognition methods detect whether a given sequence fits one of the known folds, in principle without taking into account sequence information, i.e. without assuming that the two proteins sharing the same fold, the target and the template, are evolutionary related. As already mentioned, the two proteins might be homologous, but so evolutionarily distant that their sequence similarity has dropped below the detection level, i.e. has become comparable with that expected by chance alone.

When the final model has been built, one can ask whether the residues conserved between the target and the template are located in positions crucial for their function or structure and use this information to formulate hypotheses about their evolutionary relationship. This is an important step, because detection of an evolutionary relationship might aid functional assignment of the unknown protein. Some fold-recognition methods do indeed take this possibility into account and use the sequence information to guide both the recognition of the fold and the alignment.

In general, fold-recognition methods can nowadays be regarded as quite reliable (Figure 5.5) although, in contrast with comparative modeling, it is difficult to establish a priori the expected accuracy of the model that they produce and, especially, there is no guarantee that the prediction of functionally important regions is more accurate than that of the rest of the structure, especially when the selected fold is analogous and not homologous to the query protein.

Suggested Reading

The observation that the number of folds is limited was put forward by Cyrus Chothia in:
C. Chothia (**1992**) Proteins. One thousand families for the molecular biologist. Nature **357**, 543–544

David Eisenberg first proposed that profile based methods could be used to predict protein structures:
J.U. Bowie, R. Luthy, D. Eisenberg (**1991**) A method to identify protein sequences that fold into a known three-dimensional structure. Science **253**, 164–170

The threading idea was described in:
D.T. Jones, W. R. Taylor, J. M. Thornton (**1992**) A new approach to protein fold recognition. Nature **358**, 86–89

Programs for fold recognition can be found in several sites, for example:
Threader: http://bioinf.cs.ucl.ac.uk/threader/threader.html
3DPSSM: http://www.sbg.bio.ic.ac.uk/~3dpssm/
DOE: http://fold.doe-mbi.ucla.edu/
123D: http://123d.ncifcrf.gov/
SAM: http://www.cse.ucsc.edu/research/compbio/HMM-apps/
FFAS: http://ffas.ljcrf.edu/ffas-cgi/cgi/ffas.pl
Robetta: http://robetta.bakerlab.org/

The quality of fold recognition severs is automatically assessed by Livebench and Eva:
http://bioinfo.pl/Meta/evaluation.html and http://cubic.bioc.columbia.edu/eva

6
Methods Used to Predict New Folds: Fragment-based Methods

6.1
Introduction

If the amino acid sequence of a protein does not reveal any evolutionary relationship with proteins of known structure and no fold recognition method proposes a putative fold with a sufficient level of confidence, we are left with the problem of predicting the structure *ex novo*. It is likely, if all the previous methods have been correctly explored, that the protein has a novel, not yet observed, fold and, therefore, we cannot use any of the topologies of proteins of known structure as a template for it. The possibility of predicting the structure of a protein in the absence of a suitable template has remained elusive for a long time. Although the chance of finding templates will increase as more structures are solved, the fraction of occasions where neither comparative modeling nor fold recognition can be applied, is still sizeable. When a new fold is discovered, it is often observed that it is composed of common structural motifs at the fragment or supersecondary structural level. This prompted the development of methods, known under the name of "fragment-based", that try and use fragments of proteins of known structure to reconstruct the complete structure of a target protein.

The relationship between local sequence and local structure in proteins is highly degenerate, and fragments with identical sequence can assume completely different structures in different proteins because of the effect of long-range tertiary interactions. A method that would simply search a database of known structures for fragments with the same sequence as those in the target protein and join these fragments would not work; we cannot, therefore, simply "fish" in the database of proteins of known structure for suitable fragments. Sequence-dependent local interactions might, however, bias the local structures that can be assumed by short segments of the chain and we can expect, for reasons similar to those described when we discussed pair potentials, that the observed distribution of conformations for a fragment can be used to derive the preference of the fragment for each conformation. In other words, the distribution of conformations sampled for a local segment of the polypeptide chain can be reasonably well approximated by the distribution of structures adopted by that sequence and by closely related sequences in known protein structures.

Protein Structure Prediction. Edited by Anna Tramontano
Copyright © 2006 WILEY-VCH Verlag GmbH & Co. KGaA, Weinheim
ISBN: 3-527-31167-X

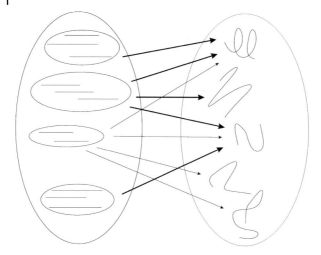

Figure 6.1 The local sequence-to-structure relationship is degenerate. For some recurring local structures, however, a correlation can be identified. In the figure, the left part depicts the space of sequence fragments. Each line represents a sequence and two similar sequences are closer to each other. Some groups of sequences will show preference for a subset of local structures (indicated by the thicker lines in the figure) while others will be less specific.

Previous studies, based on tabulations of statistics on sequences found in a given structural motif did not reveal a sufficiently significant relationship between local sequence and local structure. For example, it is well known that identical pentapeptides can be found in completely unrelated conformations. The inverse approach is more successful – by considering recurrent structural motifs and clustering them is it possible to identify sequence patterns for which a single local structure predominates. In other words, the mapping between local sequences and some recurrent local structures, such as helices, helix ends, and turns is less degenerate than for generic structural fragments (Figure 6.1).

We can use this observation to narrow the search of conformational space by preselecting structural fragments from a library of solved protein structures.

6.2
Fragment-based Methods

The overall strategy adopted by fragment-based methods is to collect the local structures assumed by short sequence segments in known three-dimensional structures and use their combinations to produce a large number of putative three-dimensional models for the target protein, among which the final model is selected on the basis of energy considerations. An important aspect of these methods is therefore the form and terms of the function used to discriminate between the many different assemblies compatible with the distribution of local

Figure 6.2 Some examples of fragment-based predictions submitted to CASP experiments.

structures. The two most popular ones are, undoubtedly, Rosetta, developed by David Baker's group, and Fragfold, proposed by David Jones. On several occasions both methods have been shown to produce impressive models of proteins with novel folds. The success of these methods has been very important in the field, because, although their accuracy does not reach that of the methods that we described in previous chapters (see Figure 6.2 for some examples), they are the only route to the prediction of proteins with novel folds.

Given the complete sequence of our protein, they:
- split the sequence into fragments;
- for each fragment, search a database of known structure for regions with a similar sequence, called neighbors; and
- use an optimization technique to find the best combination of fragments.

6.3
Splitting the Sequence into Fragments and Selecting Fragments from the Database

There are different options for splitting the sequence into fragments. Rosetta uses nine-residue-long fragments, on the basis of a study showing that the correlation between the local sequence and the local structure is greater for fragments of this length than for other fragment lengths of less than 15 amino acids.

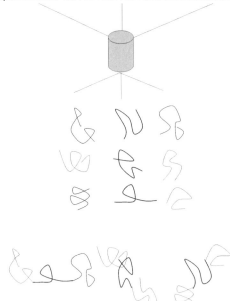

Figure 6.3 Schematic explanation of the first steps of the Rosetta method. The query sequence is split in fragments nine amino acids long. Each fragment sequence is used to search for similar fragments among the sequences of proteins of known structure. Next, the fragments are joined.

The sequence of these fragments is used to search the database of known structures for the closest 25 "neighbors" (Figure 6.3). The distance *dist* between two sequences is defined as:

$$dist = \sum_{1}^{9} \sum_{1}^{20} S(aa,i) - X(aa,i)$$

If multiple sequence alignments are available for the target and/or the template, $S(aa,i)$ and $X(aa,i)$ represent the frequencies of the amino acid aa in position i in the two alignments, respectively. If this is not true, the element of the S and X vector corresponding to the amino acid present in position i is set to unity and all the others to 0 (Figure 6.4).

The Fragfold method instead selects favorable supersecondary structural fragments at each residue position along the target sequence, in particular it uses alpha-hairpins and alpha corners (consecutive alpha helices in a compact or non-compact arrangement), beta hairpin and beta corners (consecutive beta strands hydrogen-bonded or not hydrogen-bonded to each other), beta–alpha–beta units that were described in Chapter 1 and split beta–alpha–beta units (two non hydrogen-bonded beta strands with an intervening alpha helix). It also uses a fragment list constructed from all tripeptide, tetrapeptide and pentapeptide fragments from the database of known structures.

For each position in the sequence, the method folds the fragment as each of the selected supersecondary structure fragments and computes the knowledge-based potential of the fragment. It then selects a predefined number of low-energy alternatives (for example ten). In one implementation, Fragfold excludes from this list all fragments for which the predicted secondary structure of the target sequence does not match the secondary structure of the fragment.

The optimization technique requires that we define the function to be optimized, this can be the knowledge-based potential described in the context of fold-recognition methods, but other routes are possible. We seek the most probable structure for a protein given its amino acid sequence and examples of sequences with known structures in the protein database. This problem can be addressed using Bayesian logics, a branch of logic that deals with probability inference, i.e. describes how to use the knowledge of prior events to predict future events. Using Bayes theorem, the probability of a structure given the amino acid sequence (and the information in the protein database) is:

P(structure|sequence) = P(structure) × P(sequence|structure) / P(sequence)

where P(structure|sequence) is the probability of observing a structure given a sequence, P(structure) is the prior probability of observing the structure, P(sequence|structure) is the probability of observing the sequence given the structure and P(sequence) the probability of observing the sequence.

Target   Template
Alignment   sequence

AGCTAVTAR   VGCASVTAK
VGCSTFSAK
AGCTVVATK

A 200010120   A 000100010
C 003000000   C 001000000
D 000000000   D 000000000
E 000000000   E 000000000
F 000001000   F 000000000
G 030000000   G 010000000
H 000000000   H 000000000
I 000000000   I 000000000
K 000000002   K 000000001
L 000000000   L 000000000
M 000000000   M 000000000
N 000000000   N 000000000
P 000000000   P 000000000
Q 000000000   Q 000000000
R 000000001   R 000000000
S 000200100   S 000010000
T 000110110   T 000000100
V 100012000   V 100001000
Y 000000000   Y 000000000
W 000000000   W 000000000

$Dist_1 = |2/3 - 0| + |1/3 - 1| = 4/3$
$Dist_2 = |1 - 1| = 0$
$Dist_3 = |1 - 1| = 0$
$Dist_4 = |2/3 - 0| + |1/3 - 0| + |0 - 1| = 2$
$Dist_5 = |1/3 - 0| + |1/3 - 0| + |1/3 - 0| + |0 - 1| = 2$
$Dist_6 = |2/3 - 1| + |1/3 - 0| = 2/3$
$Dist_7 = |1/3 - 0| + |1/3 - 1| + |1/3 - 0| = 4/3$
$Dist_8 = |2/3 - 1| + |1/3 - 0| = 2/3$
$Dist_9 = |2/3 - 1| + |1/3 - 0| = 2/3$
$Dist = 4/3 + 0 + 0 + 2 + 2 + 2/3 + 4/3 + 2/3 + 2/3 = 8.67$

Figure 6.4 Calculation of the distance between the sequence of a fragment of a query protein and that of a fragment of a protein of known structure, as implemented in the Rosetta method. In the example, a multiple sequence alignment is available for the query sequence and this enables a profile to be derived for each of the nine positions. The fragment of the database in the example is instead unique and its profile only contains 1 in the row corresponding to the observed amino acid and 0 in all other cells of the matrix. For each position, the distance is computed as the absolute value of the difference between the frequency of each amino acid in the profiles of the query and database sequences. They are summed to give the distance between the two sequences.

6.3 Splitting the Sequence into Fragments and Selecting Fragments from the Database

We will use an example to explain Bayesian logics. Let us assume that we have reasons to believe (from past experience) that there is a probability of 999/1 000 that a coin is fair (i.e. that there is a 50 % chance of obtaining head when we flip it) and 1/1 000 that we always get head. The probability 999/1 000 is called prior probability.

If we now flip the coin a few times, we can use the new data to re-estimate the probability that our hypothesis is true.

Let us call H1 and H2 the hypotheses that the coin is fair and that it is not, respectively. Before flipping the coin, we have P(H1) = 999/1 000 and P(H2) = 1/1 000.

The probability that the coin is fair and the outcome is a given set of flips D is:

$$P(D\ \&\ H1) = p(D)\ p(H1|D) \qquad (1)$$

i.e. the probability of observing the data D (a set of flip outcomes) in the hypothesis H1 is the probability of observing the data times the probability that H1 is true, given the new data.

P(D & H1) is also equal to:

$$P(D\ \&\ H1) = p(H1)\ p(D|H1) \qquad (2)$$

i.e. the probability of observing the data D (a set of flip outcomes) in the hypothesis H1 is the probability that the H1 is true times the probability to observe the data in the hypothesis that H1 is true. By combining equations (1) and (2) we obtain:

$$P(D|H1) = p(D)\ p(H1|D)/p(H1) \qquad (3)$$

And, analogously:

$$P(D\,H2) = p(D)\ p(H2|D)/p(H2) \qquad (4)$$

In other words, the posterior (after the observation) probability of our hypothesis is the product of the prior probability of the hypothesis times the probability of observing the new data, if the hypothesis is correct, divided by the probability of observing the new data:

Posterior = prior × (prob (new data|hypothesis H1 is true) / prob (new data)

Combining (3) and (4):

$$P(H1|D)/P(H2|D) = P(D|H1)/P(D|H2) \times P(H1)/P(H2)$$

If we now flip the coin and obtain five heads, P(D|H1), the probability of obtaining the observed data in the hypothesis that the coin is fair, is equal to $1/2^5$ and

P(D|H2), the probability of obtaining the observed data in the hypothesis that the coin is not fair, is equal to 1. Our prior probability tells us that:

P(H1)/P(H2) = (999/1000)/(1/1000) = 999

P(H1|D)/P(H2|D) = $1/2^5$ × 999 = 31%.

We can apply Bayes logics in many fields for testing hypotheses and derive posterior probabilities of events given prior probabilities and new data.

In Rosetta-like fragment-based methods we divide our protein into fragments of a given length and search for each of the fragments in a database of known protein structures to determine the probability that the fragment is found in a certain conformation. This implies that, each time, we search for a given sequence, therefore P(sequence) is always 1:

P(structure|sequence) = P(structure) × P(sequence|structure)

Now we need to compute the prior probability P(structure) of the structure. We can assume that each structure is equally probable for our sequence and set P(structure) to 1/(Number of structures). It is, however, also useful to include extra terms since we know that low-energy folds are compact, have optimum hydrogen-bond networks, and have no steric clashes. In fold recognition, these additional terms are unnecessary, because real protein folds are almost always compact, have no steric clashes, and have well-defined hydrogen-bonding networks.

Different methods, and different implementations of the same method, use different expressions for P(structure) trying to take these factors into account. For example, one could use P(structure) = 0 for fragments with overlapping atoms and assign a P(structure) related to the radius of gyration to all other fragments, or we may take into account the orientation of local structure elements by relating P(structure) to the separation and relative orientation of local structural elements.

Question: What is the radius of gyration?

»The radius of gyration is a property characterizing the size of a particle of any shape. For a rigid particle consisting of mass elements of mass m_i, each located at a distance r_i from the center of mass, the radius of gyration, g, is defined as the square root of the mass-average of r_{i2} for all the mass elements, i.e.:

$$g = \left(\frac{\sum_i m_i r_i^2}{\sum_i m_i} \right)^2 \text{«}$$

The term p(sequence|structure) is the probability of observing a conformation given the sequence of the fragment. We can use terms related to the solvent accessibility and secondary structure preference to evaluate the likelihood that a given sequence assumes the structure we are observing. Taking into account what we said so far, we can compute the probability that a fragment assumes a certain conformation in our final protein structure, i.e. P(structure|sequence).

6.4
Generation of Structures

To generate the set of structures for the target protein, Rosetta uses simulated annealing. It starts from an extended chain and each move consists in substituting the dihedral angles of a randomly chosen neighbor at a randomly chosen position for those of the current configuration. Conformations are initially evaluated using Bayesian derived probabilities. In subsequent cycles, knowledge based potentials are used.

Fragfold generates a random conformation for the target sequence by selecting fragments entirely randomly. Fragments are joined by superposing the α-carbon and the main-chain nitrogen and carbonyl-carbon atoms of the C-terminus of one fragment on the equivalent atoms of the N-terminus of the other fragment. If the resulting conformation has any pair of atoms closer than a predetermined minimum distance, it is rejected and another randomly generated conformation is selected by using the same procedure. This is repeated until the starting conformation has no steric clashes. When a conformation with no clashes has been obtained, the conformation is modified by randomly selecting a fragment conformation from the fragment lists. The choice is biased by the value of the knowledge-based potential of the fragments (low-energy fragments are more likely to be selected). Fragfold also uses simulated annealing. Each move is made by either selecting a supersecondary structure fragment, or a completely free choice is made from the additional list of small fragments. Half of the moves made involve a supersecondary structure fragment, and half involve a free selection from the small fragment list.

In general in these methods, several simulations are run for each target sequence and the results from one is chosen as the final model according to its energy. Sometimes the final results are clustered and the size of the cluster is used as an additional criterion for selecting the model.

The development of fragment-based methods is undoubtedly the most exciting new advance of the past few years in protein structure prediction and we are just starting to see the effect these methods are having in the biological sciences. So far, they are still rather computer intensive and this is limiting their usage. Automatic prediction servers using fragment-based methods are still few and take a long time to provide an answer, and this sometimes discourages users. They are also having an impact on structure prediction in other subject areas, for example to attempt prediction of structurally divergent regions in comparative models.

The Rosetta method has also been applied to de novo protein design. Protein design is the inverse of the folding problem. The challenge in this case is to find a sequence able to fold into a given structure. The structure can be that of an existing protein, so that the problem is to redesign its sequence, or a completely novel fold, not yet observed. Scientists have been working on this problem for more than a decade, with variable results. Recently, in a very successful experiment, the Rosetta method was used to design an artificial sequence able to fold into a structure with a topology not yet observed in any natural protein. The impressive similarity between the designed and experimental structure, subsequently determined, shows that the goal of designing customized proteins able to fold in a predetermined way, and possibly performing a desired function, is within our reach.

Suggested Reading

The two fragment based methods, Rosetta and Fragfold, are described in:
K.T. Simons, R. Bonneau, I. Ruczinski, D. Baker (**1999**) Ab initio protein structure prediction of CASP III targets using ROSETTA Proteins Suppl **3**, 171–176
D.T. Jones (**1997**) Successful ab initio prediction of the tertiary structure of NK-lysin using multiple sequences and recognized supersecondary structural motifs. Proteins Suppl **1**, 185–191

Each of the CASP issues (published every two years in the journal Proteins: Structure, Function and Bioinformatics published by Wiley) contains a section dedicated to the results of these methods.
Readers interested in protein design can also read:
B. Kuhlman, G. Dantas, G. C. Ireton, G. Varani, B. L. Stoddard, D. Baker (**2003**) Design of a novel globular protein fold with atomic-level accuracy. Science **302**, 1364–1368

7
Low-dimensionality Prediction: Secondary Structure and Contact Prediction

7.1
Introduction

The local structure of a protein can be described in terms of its secondary structure, i.e. of the location of alpha helices and beta strands, as already discussed. The secondary structure of a protein can also be encoded by a linear string of characters, for example H for alpha helices, E for beta strands, and L for all other regions. Prediction of the location of secondary structure elements in a protein can therefore be described as a mapping problem in which we need to relate a string encoded by an alphabet of twenty letters (the sequence) into a string using an alphabet of three characters (Fig. 2.1), or a few more if we want to distinguish between different types of helix and different types of turn. This way of posing the problem widens the range of algorithms that can be used for prediction. For example, we can use automatic learning methods, i.e. methods that try to infer the relationships between objects by learning them from a set of known cases. In practice, the methods try to identify common features of the input values that are associated with the same output values.

The three-dimensional representation of protein structures, i.e. the list of their 3D coordinates, is not a good representation for automatic learning approaches, because similar features in different proteins have completely different "values" (coordinates), because each protein has its own coordinate system. The secondary structure, instead, is an ideal representation for mapping methods and, indeed, the simplicity of the formulation of this problem has attracted much attention since the early days of structural biology, when only a few protein structures were known experimentally.

The simplest hypothesis is that amino acids have a preference for one or other of the secondary structure elements and that these preferences can be used to predict the location in the sequence of helical and beta segments. We will shortly review the history of methods based on this hypothesis, because they are instructive examples of strategies used to extract information from experimental observation of protein structures. Before doing that, however, we should ask ourselves a basic question – suppose we could predict with infinite accuracy the secondary structure of a protein,

Protein Structure Prediction. Edited by Anna Tramontano
Copyright © 2006 WILEY-VCH Verlag GmbH & Co. KGaA, Weinheim
ISBN: 3-527-31167-X

how much would that help in reconstructing the complete structure of the protein? The problem is not trivial and, indeed, in general, the location of secondary structure elements does not uniquely define the three-dimensional structure of a protein. Other information is needed, for example some constraints on their relative distance. In the last part of this chapter we will discuss methods that attempt to derive distance constraints between amino acids on the basis of a multiple sequence alignment of proteins of a family. They can help positioning the various parts of the protein. Should this be found to be an effective means of obtaining a sufficient number of reasonably accurate upper and lower bounds on the distance between elements of secondary structure, we could attempt reconstruction of the complete conformation of a target protein, much in the same way as we do when reconstructing a three-dimensional structure from NMR data (Chapter 1). We are not there yet, but it is encouraging to see that a first step has been made – the accuracy of methods for secondary structure prediction has reached very respectable values, approximately 80%.

Question: How can we reconstruct a protein structure from a set of distances between its atoms and how many distance constraints do we need?

»In a protein with N atoms, there are $N \times (N-1)/2$ distances we can potentially measure and $3N$ coordinates we need to compute. Furthermore, the geometry of proteins adds more constraints to the problem (for example we know how atoms are connected in an amino acid and how different amino acids are connected together). Therefore, in principle, if we know a sizeable fraction of the values of the distances between the elements, the complete three-dimensional structure can be reconstructed. Even in ideal NMR experiments, however, and more so in computational methods, not all atomic distances can be measured and, what is more important, the uncertainty in the values of the distances is rather high, and this makes the problem complex.«

The method most often used for reconstructing a protein structure from a set of distances is "distance geometry". The method relies on the observation that the scalar product g_{ij} of the vectors $\mathbf{x}_i$ and $\mathbf{x}_j$ representing the coordinates of two atoms is equal to:

$$g_{ij} = \tfrac{1}{2}(d^2_{io} + d^2_{jo} - d^2_{ij})$$

where d_{io} and d_{jo} are the distances of atoms i and j, respectively, from a chosen origin of the coordinates and d_{ij} is the distance between atoms i and j. It can be demonstrated that, if we select the origin of the coordinates as the centroid of all atoms:

$$x_{ik} = \lambda_k^{1/2} w_{ik}$$

where λ_k and w are the eigenvectors and eigenvalues of the matrix g.

7.1 Introduction

Let us see what they are. Given a square matrix g (i.e. a matrix with the same number of columns and rows), we say that λ is an eigenvalue of A if there exists a non-zero vector X such that $AX = \lambda X$. In this case, X is called an eigenvector (corresponding to λ), and the pair (λ, X) is called an eigenpair for A. There can be more than one eigenpair for a given matrix, as shown in Figure 7.1.

$$A = \begin{bmatrix} 13 & -4 \\ -4 & 7 \end{bmatrix} \quad x = \begin{bmatrix} -2 \\ 1 \end{bmatrix} \quad \lambda = 15$$

$$A = \begin{bmatrix} 13 & -4 \\ -4 & 7 \end{bmatrix} \begin{bmatrix} -2 \\ 1 \end{bmatrix} = \begin{bmatrix} (13)(-2) + (-4)(1) \\ (-4)(-2) + (7)(1) \end{bmatrix} = \begin{bmatrix} -30 \\ 15 \end{bmatrix} = \lambda \begin{bmatrix} -2 \\ 1 \end{bmatrix} ; \lambda = 15$$

$$A = \begin{bmatrix} 13 & -4 \\ -4 & 7 \end{bmatrix} \quad x = \begin{bmatrix} 1 \\ 2 \end{bmatrix} \quad \lambda = 5$$

$$A = \begin{bmatrix} 13 & -4 \\ -4 & 7 \end{bmatrix} \begin{bmatrix} 1 \\ 2 \end{bmatrix} = \begin{bmatrix} (13)(1) + (-4)(2) \\ (-4)(1) + (7)(2) \end{bmatrix} = \begin{bmatrix} 5 \\ 10 \end{bmatrix} = 5 \begin{bmatrix} 1 \\ 2 \end{bmatrix}$$

Figure 7.1 The matrix **A** has two pairs of eigenvectors and eigenvalues. The first pair is $(-2,1)$ and 15, the second $(1,2)$ and 5.

Eigenpairs are extremely useful in a large number of applications, from engineering to calculus. You can think of them as a way to decompose and project the distance matrix to obtain vectors representing the coordinate axes of a new reference system.

If the distances have been computed for a real three-dimensional object and they have no errors, only the first three eigenvectors are different from zero. When the matrix is not complete (i.e. some distances are not known) and distances contain errors, we can use some approximations. For example, if the cell corresponding to the distance between atoms k and l is empty, but we know the distances d_{ki} and d_{li} of our atoms from a third atom i, we can use the triangle inequalities $d_{kl} < d_{ki} + d_{li}$ and $d_{kl} > |d_{ki} - d_{li}|$ (Figure 7.2) and approximate the value of the (k,l) cell to, for example, the average between $d_{ki} + d_{lj}$ and $d_{ki} - d_{lj}$. If more than three eigenvalues are different from zero, we can use the three largest eigenvalues of the matrix to compute the coordinates.

It follows that the number of distances that we need is correlated to the sparseness (i.e. number of unfilled cells) of the matrix and on the uncertainty in the distances.

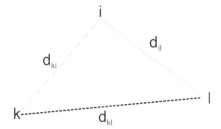

Figure 7.2 Illustration of the triangle inequalities. The distance between the points labeled **k** and **l** is shorter than the sum of the d_{ki} and d_{il} distances and longer than their difference.

7.2
A Short History of Secondary structure Prediction Methods

As already mentioned, the history of secondary structure prediction methods dates back some time. It started as soon as the first protein structure determinations were available. In early approaches, frequencies of occurrence of particular amino acids in each of the secondary structures were used to compute the probability a given segment was in one of the secondary structure conformations. Empirical rules were usually used to obtain the final prediction. As an example we will describe in some detail the very popular Chou and Fasman algorithm, not because it is advisable to use it today (much more accurate methods are available), but because it is part of the "history" of bioinformatics and is often mentioned as an example of a knowledge-based prediction method.

The data set available at the time was very limited, only 29 protein structures were known. Chou and Fasman used it to compute, for each of the twenty amino acids, how many times they occurred in alpha helices, strands, and coil in their collection of proteins. The logarithm of these numbers divided by the expected counts (i.e. the expected frequency for each amino acid if it was equivalently distributed in the three classes) to which the value 1.0 was added, are the "preference parameters", called Pa, Pb, and Pc for alpha helices, beta strands, and coil, respectively (Table 7.1).

Table 7.1 The Chou and Fasman parameters for secondary structure prediction.

Amino acid	Pa	Pb	Pc	Amino acid	Pa	Pb	Pc
Alanine	1.42	0.83	0.66	Leucine	1.21	1.30	0.59
Arginine	0.98	0.93	0.95	Lysine	1.14	0.74	1.01
Aspartic acid	1.01	0.54	1.46	Methionine	1.45	1.05	0.60
Asparagine	0.67	0.89	1.56	Phenylalanine	1.13	1.38	0.60
Cysteine	0.70	1.19	1.19	Proline	0.57	0.55	1.52
Glutamic acid	1.39	1.17	0.74	Serine	0.77	0.75	1.43
Glutamine	1.11	1.10	0.98	Threonine	0.83	1.19	0.96
Glycine	0.57	0.75	1.56	Tryptophan	1.08	1.37	0.96
Histidine	1.00	0.87	0.95	Tyrosine	0.69	1.47	1.14
Isoleucine	1.08	1.60	0.47	Valine	1.06	1.70	0.50

If an amino acid is observed in a secondary structure element as often as expected by chance, the ratio between the observed and expected frequency is 1, the logarithm of 1 is 0, and, consequently, its preference parameter for the specific secondary structure is 1. For example, Pb for isoleucine, a beta branched amino acid, is 1.60, therefore the logarithm of the ratio between the frequency of this amino acid in alpha helices relative to that expected by chance is 0.60. This implies

	P	T	L	E	W	F	L	S	H	C	H	I	H	K	Y
Pa	0.57	0.83	1.21	1.39	1.08	1.13	1.21	0.77	1.00	0.70	1.00	1.08	1.00	1.14	0.69
Pb	0.55	1.19	1.3	1.17	1.37	1.38	1.30	0.75	0.87	1.19	0.87	1.6	0.87	0.74	1.47
Predicted	E	E	E	E	E	E	E								
Observed	H	H	H	H	H	H	H	H	H					E	E

	P	S	K	S	T	L	I	H	Q	G	E	K	A	E	T
Pa	0.57	0.77	1.14	0.77	0.83	1.21	1.08	1.00	1.11	0.57	1.39	1.14	1.42	1.39	0.83
Pb	0.55	0.75	0.74	0.75	1.19	1.30	1.60	0.87	1.10	0.75	1.17	0.74	0.83	1.17	1.19
Predicted			E	E	E	E	E		E	E	E	E	E	E	E
Observed															E

	L	Y	Y	I	V	K	G	S	V	A	V	L
Pa	1.21	0.69	0.69	1.08	1.06	1.14	0.57	0.77	1.06	1.42	1.06	1.21
Pb	1.30	1.47	1.47	1.60	1.70	0.74	0.75	0.75	1.70	0.83	1.70	1.30
Predicted	E	E	E	E		E	E	E	E	E	E	
Observed	E	E	E					E	E	E	E	E

Figure 7.3 An example of the application of the Chou and Fasman method to a protein. It is apparent that the predicted and experimental structures only have limited overlap.

that isoleucine is observed in beta strands approximately 1.8 ($e^{0.6}$) times more often than expected by chance.

Given the alpha, beta and coil parameters, the method proceeds through three steps. In the first step the amino acid sequence is converted into three strings of numbers, one in which each amino acid is replaced by its alpha parameter, one in which it is replaced by its beta parameter and one in which it is replaced by the coil parameter (Figure 7.3).

In the second step the algorithm searches for regions where 4 out of 6 contiguous residues have *Pa* higher than 1.0. The region is temporarily defined as a helix and is subsequently extended in both directions until the average *Pa* value computed for a set of four contiguous residues is lower than 1.0. If, at the end of this step, the segment is longer than five residues and the average *Pa* is higher than the average *Pb* for the whole region, then the segment is predicted to be helical.

A similar procedure is used to identify putative beta strand nucleation sites. Three out of five contiguous residues must have a *Pb* above 1.00 and the segment is extended in both directions until the average *Pb* value of four consecutive residues is below 1.0. If, at the end of this step, the average *Pb* is higher than the average *Pa* for the whole region, it is defined as a beta segment. If, after this procedure, a region is predicted to be both helical and beta, the final decision depends on which average preference value, *Pa* or *Pb*, is higher (Figure 7.3). A slightly more complex rule is used to identify turns, but also in this case the prediction is based on the observed frequency of residues in this type of local structure.

The hypothesis underlying this and other, similar, methods, is that local interactions dominate in determining whether or not a region assumes a given secondary structure. These methods also implicitly assume that there are "nucleation sites" for secondary structure, so that the polypeptide chains starts folding by forming local structures in regions composed of amino acids with very high tendency to form one of the secondary structure types. Subsequently these structures extend toward the amino and carboxy termini of the protein until a region with low propensity to form the secondary structure is encountered. Eventually the secondary structure elements collapse against each other to give the native structure.

Other methods followed, for example the GOR (Garnier Osguthorpe Robson) method. These authors used a set of known three-dimensional structures of proteins to compute three tables listing the frequency of stretches of amino acids seventeen residue long in which the central amino acid is in an alpha helical (fα), beta (fβ), or coil (fc) conformation. Given a new sequence, for each amino acid, i, one computes the values:

$$P\alpha_i = \sum_{j=i-8}^{j=i+8} f\alpha_i \quad P\beta_i = \sum_{j=i-8}^{j=i+8} f\beta_i \quad Pc_i = \sum_{j=i-8}^{j=i+8} fc_i$$

and assigns the amino acid to the alpha, beta, or coil class according to which of the three values is higher. When careful analysis of these methods was conducted by independent researchers it was discovered that their levels of accuracy were rather low, approximately 57% for the Chou and Fasman method and approximately 60–65% for the others.

The major advance came when two strategies, automatic learning methods and evolutionary information, were combined.

7.3
Automatic learning Methods

7.3.1
Artificial Neural Networks

Artificial neural networks are well suited to mapping problems such as that of secondary structure prediction. In its simplest form, a neural network is an algorithm able to separate two sets of data on the basis of parameters "learned" by the system on the basis of a training set containing data for which the mapping

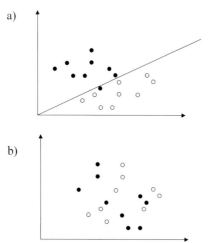

Figure 7.4 The filled and open circles represent the two sets of data that must be separated. In the example shown in (a) it is easy to find a line that separates the data with reasonable accuracy; the task is much more difficult in the example illustrated in (b).

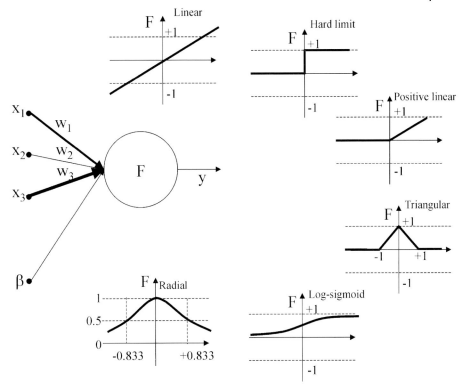

Figure 7.5 A node of a neural network is a computational unit that transforms input values x_i into an output value y. The input values can have different weights and the transformation can be different. A few examples of "transfer functions", used to compute the output of the node as a function of the weighted sum of its inputs, are shown in the figure.

is known. The general idea behind a classification method is: given a data set including points of two different types, which function discriminates better between the two sets? If the properties of the data can be identified, the data can be plotted as shown in Figure 7.4 (where we assume only two variables) and the problem is to find the line, plane, or hyperplane, depending on the number of dimensions, that best separates the two sets. The problem can be easy, as in the example shown in Fig 7.4 a, where a simple linear function is sufficient to separate the data, or more complex, as exemplified by the plot in Figure 7.4 b.

In neural networks, properties of the data are used as input to a system of nodes or "neurons". Nodes such as that shown in Figure 7.5 are the basic units of a neural network, They are computational units that receive a number of inputs. Each input comes via a connection that has a strength (or weight). Each neuron also has a single threshold value β. The weighted sum of the inputs is computed and the threshold subtracted, to calculate the activation of the neuron. The activation signal is passed through an activation or transfer function to produce the output of the neuron. In other words, given the input values $x_1, x_2, \ldots x_N$, the output of the node is:

$$Y = F(\sum_{i=1}^{N}(w_i x_i - \beta)$$

where F is the "transfer function". We can use the hard limiter (i.e. $F(x) = 1$ if x is greater than a threshold and 0 otherwise), a sigmoid function ($F(x) = 1/(1 + e^{-x})$), etc. (Figure 7.5)

A network with neurons arranged in a layered topology where information is provided as input and propagated in a forward manner is called feedforward. The input layer receives the values of the input variables. The hidden and output layer neurons are each connected to all of the units in the preceding layer (Figure 7.6).

The main feature of neural networks is that they learn the input/output relationship by training. In practice, the weights of the connections and the thresholds of the neurons are modified to maximize, for the training set, the overlap between the known answer and the output value. They can be used to find clusters in the input data (unsupervised training) or to learn the relationships between input and output using a set of known examples (supervised training). The latter is the most common and is the method used in secondary structure prediction methods.

For supervised learning we need a set of training data for which both the input and the output values are known, for example a set of sequences of proteins for which we know the three-dimensional structure and therefore also the secondary structure. These data, encoded according to some numerical scheme, are fed into the network and propagated through its layers. The output is then compared with the known desired results and the network's weights and thresholds are adjusted to minimize the errors in its predictions. If the network is properly trained it learns to model the (unknown) function that relates the input variables to the output variables and can subsequently be used to make predictions when the output is unknown. The method most commonly used to adjust the weights is called backpropagation. It involves a forward pass that generates the output and a backward pass in which the error found in the output node is used to assign errors to each of

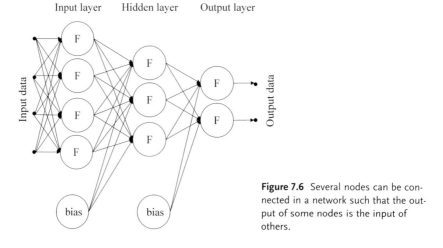

Figure 7.6 Several nodes can be connected in a network such that the output of some nodes is the input of others.

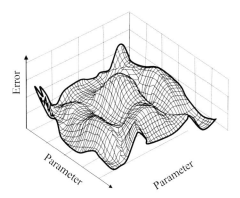

Figure 7.7 A schematic representation of an error surface. In reality the dimensions of the space is equal to the number of weights and biases in the network plus one. The task is to find the global minimum in the error surface.

the nodes. The weights are then adjusted to minimize these errors. One way of envisaging this process it to consider each individual weight and threshold as a dimension in space. If we could plot the value of the error for each combination of weights and thresholds, we would obtain an "error surface" in multidimensional space (Figure 7.7). The objective is to find the lowest point in this many-dimensional surface and there are algorithms for doing so, although none can guarantee that the global minimum is reached. When the weights have been optimized, they

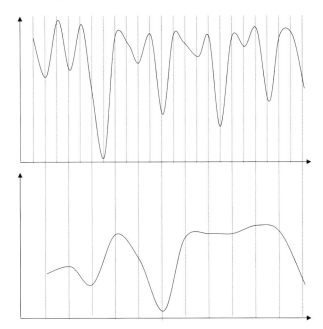

Figure 7.8 Illustration of the concept of sampling. The top panel shows a hypothetical error surface (with only two dimensions). The lines indicate the points that have been sampled, i.e. the values of the parameter for which the function has been computed. In the lower panel, the sampling is more sparse. The "surface" is approximated using a lower number of points and its details are reconstructed with lower accuracy.

are fixed. New data, with unknown output, can be input into the network to obtain a prediction.

Question: How many known examples do we need to train a neural network?

»There are some rules of thumb, for example that there should be ten times more examples than there are connections in the network, but these rules should not be taken too strictly because, clearly, the amount of data needed depends on the complexity of the problem. Each additional weight and threshold in a network adds dimensions to the error space and there must be sufficient data points to be able to describe its structure and find the global minimum (Figure 7.8). The number of points needed grows very rapidly with the dimensionality of the space. This behavior of the network, referred to as the "curse of dimensionality" is a problem, and one should try to reduce as much as possible the number of variables by eliminating unnecessary ones.«

As we said, we expect that the weights learned during the training phase of the network are the correct ones to predict the output when new cases are input into the network. In other words we would like the network to be able to generalize what it has learned to new cases, although it has only been trained to minimize the errors on the training set. In networks, as in all fitting methods, there is the risk of over-training – the network might learn very well how to reproduce the output for the examples in the training set, but be unable to generalize. The problem is not different to that encountered when trying to fit a polynomial curve to a set of data. As data are noisy, we do not expect the fitting curve to pass exactly through the points. A low-order polynomial may not be sufficient to fit the data, but if we increase the power of the polynomial (i.e. the number of variables) too much, we will fit the data exactly with a curve that might be unrelated to the underlying process. The curve might then not reflect the relationship that we are attempting to discover and become useless for predicting where our next data point is likely to fall (Figure 7.9).

This implies that selecting the complexity of the network is not easy: increasing it will reduce the error, but it might also reduce its predictive power. Usually, an independent data set, not used for training, is used as validation set to verify whether the error on the weights is becoming lower during the training process, as judged by the performance of the network in cases not used for the training. Of course, if we repeat this validation step too many times, changing the number of nodes and layers, we might optimize our network for the validation set and once again not improve its performance on new data. Ideally, a different validation set should be used every time the network structure is changed. At the end of the training we need yet another dataset, independent of the training and validation set, to assess the performance of the network. Often, the number of available cases with known output is not sufficient to follow these rules strictly, but it is important to realize what we are optimizing in our procedure.

In summary, implementation of a neural network for a given task consists of the following steps:

1. Select an initial structure for the network (number of nodes, transfer function, number of layers, etc.)
2. Select a training set, a test set, and, ideally, a number of validation sets, all independent of each other
3. Train the network
4. Evaluate network performance on a validation set
5. Change the network structure
6. Iterate steps 1 to 5, using a different validation set every time, stopping when the results on the validation set do not seem to improve further
7. Select the best network structure
8. Test its accuracy on an independent test set

In neural networks developed for protein structure prediction another complication arises – evolutionarily related proteins have, usually, the same secondary structure. We have to ensure that no protein homologous to those of the training set is present in the test and validation sets, otherwise our evaluation of the accuracy of the method is bound to be incorrect. The network may "learn" to recognize homologous proteins and to give the same answer for them, rather than to recognize the features of the sequence that make it assume one or the other secondary structure. It follows that the sequences of the proteins used in the

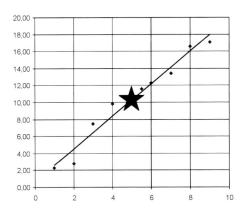 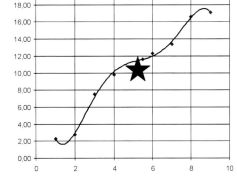

Figure 7.9 The example refers to a hypothetical experiment in which we measure the value of y for several values of x to deduce their relationship. The data for the toy experiment have been prepared using the relationship $y = x \times 2$ and adding a random error to each measurement. The best fit using a straight line is shown on the left. The resulting relationship is $y = 1.9 \times x + 0.7$. A higher-order polynomial ($y = -0.015x^5 + 0.38x^4 - 3.62x^3 + 15.25x^2 - 24.93x + 15.21$) fits the data better as shown on the right. If now we compute the value of y for $x = 5$ (the star in the figure) using the two equations we obtain $y = 10.2$ and 11.41, respectively. The higher-order polynomial of this example is overfitting the data and therefore less able to generalize.

training, validation, and test set must be compared with each other and selected in such a way that no pairs coming from different sets shares a significant sequence similarity. Usually a threshold of 25–30% sequence identity is selected for this purpose.

7.3.2
Support Vector Machines

Support vector machines are other algorithms suited to automatic classification of data. The idea is not very different from that of neural networks but it has two aspects that can sometimes make it more useful. First, the input data can be transformed using what is called a "kernel" function as to increase the probability

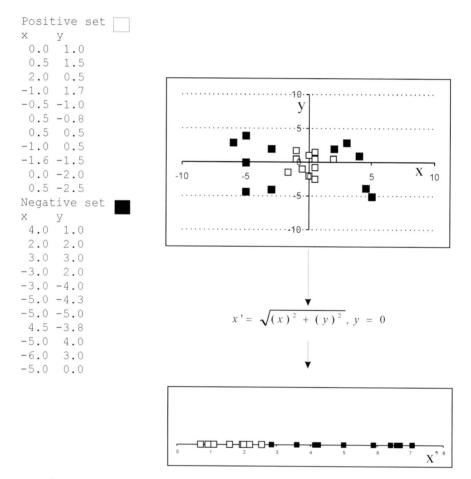

```
Positive set
x      y
 0.0   1.0
 0.5   1.5
 2.0   0.5
-1.0   1.7
-0.5  -1.0
 0.5  -0.8
 0.5   0.5
-1.0   0.5
-1.6  -1.5
 0.0  -2.0
 0.5  -2.5
Negative set
x      y
 4.0   1.0
 2.0   2.0
 3.0   3.0
-3.0   2.0
-3.0  -4.0
-5.0  -4.3
-5.0  -5.0
 4.5  -3.8
-5.0   4.0
-6.0   3.0
-5.0   0.0
```

$$x' = \sqrt{(x)^2 + (y)^2}, \; y = 0$$

Figure 7.10 In some cases, a coordinate transformation can enable separation of two sets of data originally very difficult to separate (as those shown in the top part of the figure).

that the sets of input data are, indeed, separable; this is equivalent to transforming the coordinate system of our input data in such a way that the data are easier to separate. A simple example is shown in Figure 7.10. The two types of data (white and black squares) that we aim to separate are distributed such that no plane can separate them. If, however, we transform the coordinates of our data by using the formula:

$$x = \sqrt{(x)^2 + (y)^2}, y = 0$$

i.e. by computing their distance from the origin, they are readily separable.

In practice, the support vector machine algorithm is provided with a set of putative kernel functions. These are applied to the data and the number of points that coincide in the new representation of the data is recorded. The kernel function that best separates them is ultimately chosen. This is not very computationally expensive for reasons that will not be discussed here. When the data have been transformed, we need to find the hyperplane (straight line in two dimensions) that best separates the two types of data. To minimize the risk of misclassifying unknown data, the hyperplane with the maximum distance from the closest points is chosen. One way to achieve this is shown in Figure 7.11.

Support vector machines are an extremely powerful method of obtaining models for classification. They provide a mechanism for choosing the model structure in a natural manner which gives low generalization error and empirical risk.

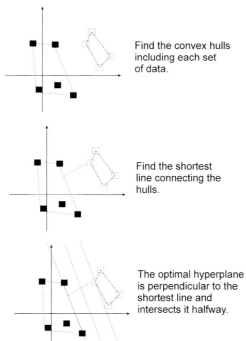

Figure 7.11 A schematic explanation of support vector machines.

7.4
Secondary structure Prediction Methods Based on Automatic Learning Techniques

How can we design a neural network or an SVM to predict secondary structure? First, we need a way to encode the amino acid sequence and the secondary structure of our proteins. We can assign a number to each of the twenty different amino acids, but we have to be careful. Numbers have relationships between them, so if we assign 1 to alanine, 2 to cysteine, and 3 to aspartic acid, we are implicitly saying that alanine and cysteine are closer to each other than alanine is to aspartic acid. We do not want to make any assumption about which amino acids are more similar with respect to their ability to assume a given secondary structure, because we do not know which properties are relevant. Therefore, we need to encode the amino acids in such a way that they are equally distant from each other. This is usually achieved by using "sparse coding", i.e. by using for each amino acid a twenty bit string with only one bit (different for each amino acid) set to unity (Figure 7.12).

	A	C	D	E	F	G	H	I	K	L	M	N	P	Q	R	S	T	V	Y	W
A	1	0	0	0	0	0	0	0	0	0	0	0	0	0	0	0	0	0	0	0
C	0	1	0	0	0	0	0	0	0	0	0	0	0	0	0	0	0	0	0	0
D	0	0	1	0	0	0	0	0	0	0	0	0	0	0	0	0	0	0	0	0
E	0	0	0	1	0	0	0	0	0	0	0	0	0	0	0	0	0	0	0	0
F	0	0	0	0	1	0	0	0	0	0	0	0	0	0	0	0	0	0	0	0
G	0	0	0	0	0	1	0	0	0	0	0	0	0	0	0	0	0	0	0	0
H	0	0	0	0	0	0	1	0	0	0	0	0	0	0	0	0	0	0	0	0
I	0	0	0	0	0	0	0	1	0	0	0	0	0	0	0	0	0	0	0	0
K	0	0	0	0	0	0	0	0	1	0	0	0	0	0	0	0	0	0	0	0
L	0	0	0	0	0	0	0	0	0	1	0	0	0	0	0	0	0	0	0	0
M	0	0	0	0	0	0	0	0	0	0	1	0	0	0	0	0	0	0	0	0
N	0	0	0	0	0	0	0	0	0	0	0	1	0	0	0	0	0	0	0	0
P	0	0	0	0	0	0	0	0	0	0	0	0	1	0	0	0	0	0	0	0
Q	0	0	0	0	0	0	0	0	0	0	0	0	1	1	0	0	0	0	0	0
R	0	0	0	0	0	0	0	0	0	0	0	0	0	0	1	0	0	0	0	0
S	0	0	0	0	0	0	0	0	0	0	0	0	0	0	0	1	0	0	0	0
T	0	0	0	0	0	0	0	0	0	0	0	0	0	0	0	0	1	0	0	0
V	0	0	0	0	0	0	0	0	0	0	0	0	0	0	0	0	0	1	0	0
Y	0	0	0	0	0	0	0	0	0	0	0	0	0	0	0	0	0	0	1	0
W	0	0	0	0	0	0	0	0	0	0	0	0	0	0	0	0	0	0	0	1

Figure 7.12 Sparse coding of amino acids. Each amino acid is represented by twenty bits, only one of which is set to 1.

7.4 Secondary structure Prediction Methods Based on Automatic Learning Techniques

Because we know that the information contained in each amino acid is insufficient to determine the local secondary structure, we might want to use more than one input node, and give to the network the binary sequence corresponding to a segment of our protein, including the N amino acids before and after that for which we want our prediction. The output nodes can be three, one for helices, one for strands, and one for coil and we can select to assign to the amino acid the secondary structure for which the output node has the highest value. Note that a network with more than one output node is more difficult to train, because a node might receive a "conflicting instruction" from the error function generated by each of the output nodes.

Training, validation and test sets, as we have discussed, must be chosen carefully. The secondary structure assignment for the training set can be derived from the three-dimensional structures of the proteins in the set using methods such as DSSP or STRIDE. The first attempts to apply neural networks to the protein secondary structure prediction problem appeared in the literature at the end of the nineteen-nineties but their performance was not significantly better than that of the preference-based methods already described. A major improvement came when, in 1994, Rost and Sander were able to demonstrate that, by using evolutionary information, the accuracy of the predictions could reach values above 70% with a method called PHD. Homologous proteins, as we have repeated several times, are expected to have similar structure and therefore also similar secondary structure. The prediction that a residue is in an alpha helical conformation in a protein should be compatible with all residues in the same position in a multiple sequence alignment of the family being in a helix. We can give this information to the neural network by using as input not a single sequence, but all the sequences of a multiple alignment of the family, encoded in the form of a profile.

This very popular PHD method is based on a rather elaborate type of neural network and uses as input a sequence profile derived from a multiple sequence alignment. Given a sequence of unknown structure, a database search for homologs through iterated PSI-BLAST is performed and the score is used to decide which proteins are to be regarded as homologs. This initial set of sequences is aligned, after eliminating redundancy if too many too similar proteins are present, and used to derive a sequence profile. A sliding window of length thirteen is moved along the profile and, at each position, the following information is fed into the network:

- the profile of amino acid substitutions for all thirteen residues, one per input node;
- the conservation weights compiled for each column of the multiple alignment;
- the number of insertions and deletions in each column;
- the position of the current segment of thirteen residues with respect to the amino and carboxy terminus of the protein;
- the amino acid composition; and
- the overall length of the protein.

The network output consists of three units, one for helix, one for strand, and one for non-regular structure (Figure 7.13). A second network uses as input the output

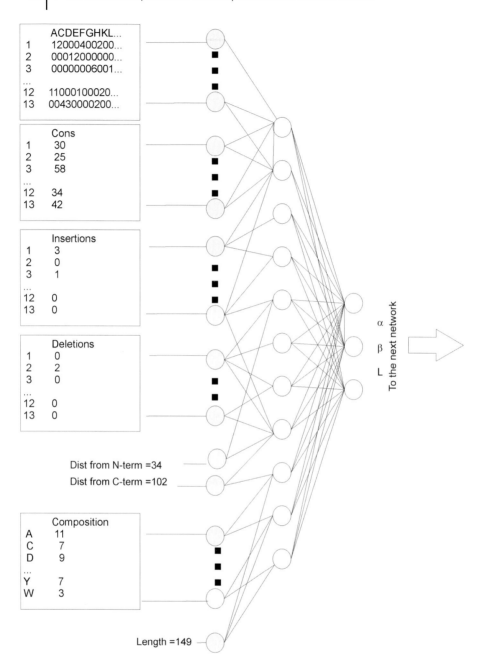

Figure 7.13 The secondary structure prediction method PHD uses several input nodes, containing information on the protein or on the thirteen-residue window under examination, to produce a prediction of the secondary structure of the central element of the window.

values of the first level network. This second network has the aim of ensuring that predicted secondary structure segments have length distributions similar to those observed in real protein structures. The third level consists of an arithmetic average over the results of independently trained networks (jury decision). The output is the prediction for the secondary structure of the central element of the window together with a value, the reliability index, that indicates how many of the independently trained networks agree on the prediction. This index is a measure of confidence in the prediction. This is an important aspect because the reported total accuracy of any method represents the average over several cases and the prediction accuracy varies with the protein. This implies that, in PHD for example, the accuracy of the prediction for a given protein can be as low as 40%, or it could be higher than 90%. The reliability index correlates with accuracy. In other words, residues with higher reliability index are predicted with higher accuracy and the reliability index offers an excellent tool to focus on some key regions predicted with high confidence.

Another reliable method of secondary-structure prediction is PSIPRED, the accuracy of which moderately exceeds that of PHD. This method also uses profiles, those generated by PSI-BLAST, as input, and is based on two feed-forward neural networks. In the prediction process it averages the output from up to four separate neural networks to increase prediction accuracy.

The list of neural network-based methods for protein structure prediction is long, and growing, as are new versions of existing methods made available to the community. As we saw in Chapter 2, selection of the appropriate method should be based on the most recent results reported by evaluation experiments such as EVA. The secondary structure prediction algorithms described here are trained with globular proteins and should not be used for transmembrane proteins and coiled coils, for which different methods, described in the next chapter, should be used.

The accuracy of all these methods for prediction of secondary structure depends on the "quality" of the multiple sequence alignment. If the protein has few homologs, or if they are all very closely related, the accuracy is likely to be lower than the reported values. Furthermore, if the protein under examination has close homologs of known structure, the secondary structure prediction implied by the sequence alignment (i.e. attributing to each amino acid the same secondary structure of the corresponding one in the protein of known structure) is more accurate than that derived from automatic learning methods.

7.5
Prediction of Long-range Contacts

If a multiple sequence alignment of members of a protein family is available, we can hope to identify residues that tend to mutate in a correlated fashion and, in these circumstances, hypothesize that the two residues are in physical contact in the protein structure. The basic hypothesis is that residues in contact in a protein structure tend to mutate in a covariant fashion – when a residue playing a crucial

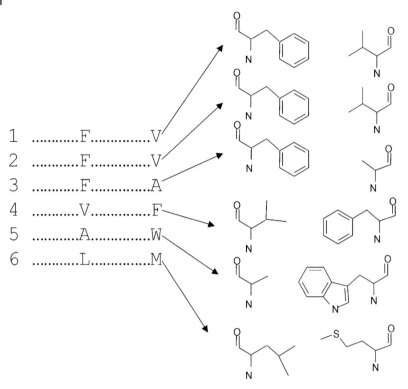

Figure 7.14 The rationale behind correlated mutations is that mutations of two amino acids in contact in a protein structure compensate each other. This is a very crude approximation because proteins behave in a more complex fashion and mutations can be accommodated because of variation in the side chain or backbone conformations, or because more than two amino acids undergo changes in the same region of the protein structure.

role in the function or structure of a protein randomly mutates, the proximal residues might be forced to compensate for the change by undergoing mutations covariant to the first. Detection of residues that mutate in a correlated fashion can therefore be taken as an indication of probable physical contact in three dimensions (Figure 7.14).

Application of this idea in RNA structure prediction has met with great success, but in proteins matters seem to be much more complex. One implementation of the method, by Olmea and coworkers, was based on the following strategy. Given a multiple sequence alignment, they constructed an $M \times N \times N$ matrix, where M is the length of the alignment and N is the number of sequences in the alignment. The $s(m,k,l)$ cell of the matrix contains the similarity between the residue observed at position m in sequence k and sequence l. The similarity values are taken from one of the BLOSUM or PAM matrices. The next step is to detect whether the elements corresponding to two positions in the alignment mutate in a correlated fashion, i.e. to compare the sub-matrices $s(i,k,l)$ and $s(j,k,l)$, where i and j are two

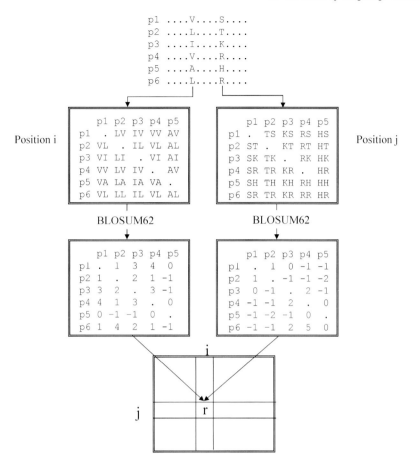

Figure 7.15 Scheme of the correlation coefficient calculation. The top shows part of a multiple sequence alignment including sequences p1, p2, ..., pN. The two matrices labeled "position i" and "position j" report the pairs observed in positions i and j of the sequences. Next the PAM62 value corresponding to the pair is computed and the correlation coefficient for the specific pair of positions is calculated.

positions in the alignment. The correlation between the behavior of the two positions i and j can be computed as:

$$r_{ij} = \frac{\sum_{kl} w_{kl}(s_{ikl} - \langle s_i \rangle) \times (s_{jikl} - \langle s_j \rangle)}{\sigma_i \cdot \sigma_j}$$

where σ_i is the standard deviation of s_{ikl} about its mean $\langle s_i \rangle$ and the indices k,l run from 1 to N, the number of sequences in the family (Figure 7.15).

The accuracy of these methods is not easy to evaluate. One must define a distance cut-off below which two residues are defined as in physical contact, and a cut-off value for the correlation coefficient above which the correlation is expected to be predictive of a contact. It is, furthermore, not easy to estimate the random expectation for such a prediction. Clearly, there is a greater probability that residues close in the sequence will be in contact, and hydrophobic residues, that tend to cluster in the hydrophobic core of the protein, are more likely to be in contact with each other than with other residues.

Subsequent refinements of the correlated mutation idea have led to methods that used the predicted contacts for construction of contact maps or combined the information with a variety of applications such as protein docking and threading. They have also been used to identify functionally important residues and residues involved in protein-protein interactions, and to predict proximal residue pairs to be used as constraints in Monte Carlo folding simulations.

Other strategies for predicting long-range contacts in protein structures from sequences are, for example, likelihood matrices. The idea is to estimate the likelihood of contact of pairs of residues of a given type from a sample of proteins of known structure. Pairs of contacts are then predicted by taking a multiple sequence alignment and summing the likelihood of all residue pairs in the corresponding columns.

Another approach to the problem has been to train neural networks or hidden Markov models using different encoding systems for multiple sequence alignments. For example, each residue pair in the protein sequence can be coded as an input vector containing 210 elements ($20 \times (20 + 1)/2$), representing all the possible ordered couples of residues (considering that each residue couple and its symmetric are coded in the same way) and a single output state can code for contact and non-contact.

Some of these methods also perform a post-processing step in which physically impossible configurations are removed from the list of predicted contacts. A possible simplification of the problems is to try and predict contacts between secondary structure elements rather than between single residues. The hope is that such interactions are more specific, and hence easier to predict. This can be done by deriving heuristic potentials for pairs of strands and/or helices and then using the potentials in conjunction with secondary structure prediction methods. Surprisingly this route has not been extensively investigated and there are very few methods that work on this basis.

The debate about the best way to evaluate accuracy of long-range contact-prediction methods is ongoing and rather technical. We will just say here that, so far, the accuracy and coverage reached by these methods is still insufficient for "folding" a protein, but it can sometimes be used to discriminate between different structural models built with one of the modeling methods already described.

Suggested Reading

The two methods for assigning the secondary structure to a protein on the basis of its coordinates are described in:

W. Kabsch, C. Sander (**1983**) Dictionary of Protein Secondary Structure: Pattern Recognition of Hydrogen-bonded and Geometrical Features **22**, 2577–2637

D. Frishman, P. Argos (**1995**) Knowledge-based protein secondary structure assignment. Proteins: Structure, Function, and Genetics **23**, 566–579

A description of the original Chou and Fasman method, based solely on the twenty-nine known protein structures available at the time, can be found in:

P.Y. Chou, G. D. Fasman (**1978**) Conformational parameter for alpha helix (computed from 29 proteins). Adv. Enzym. **47**, 145–148

Rost and Sander first published an evaluation of available secondary structure prediction methods in:

B. Rost, C. Sander (**1992**) Jury returns on structure prediction. Nature **360**, 540–540

and next proposed their own method PHD:

B. Rost, C. Sander (**1993**) Prediction of protein secondary structure at better than 70% accuracy. J. Mol Biol. **232**, 584–599

B. Rost, C. Sander (**1993**) Improved prediction of protein secondary structure by use of sequence profiles and neural networks. PNAS **90**, 7558–7562

PSI-Pred is described in:

D.T. Jones (**1999**) Protein secondary structure prediction based on position-specific scoring matrices. J. Mol. Biol. **292**, 195–202

The idea and an application of correlated mutations are described in:

U. Göbel, C. Sander, R. Schneider, A. Valencia (**1994**) Correlated mutations and residue contacts in proteins. Proteins **18**, 309–317

O. Olmea, B. Rost, A. Valencia (**1999**) Effective use of sequence correlation and conservation in fold recognition. J. Mol. Biol. **293**, 1221–1239

8
Membrane Proteins

8.1
Introduction

Prediction of the structure of membrane proteins, i.e. of proteins that are embedded in biological membranes deserves a special treatment. The roles played by membrane proteins are many, and extremely important: they can be transporters, receptors, and enzymes. These proteins are involved in many pathological processes, for example cystic fibrosis, virus entry and maturation, resistance against cytostatic drugs, bacterial infections, to name but a few. Consequently, they are among the most attractive targets for pharmaceutical intervention and there is enormous interest in predicting their structure, which depends upon the special environment in which they reside.

Membranes separate cells, or cellular compartments, from the environment and are made of lipids and protein molecules held together by non-covalent interactions. The most abundant lipids in the membrane are phospholipids, formed by a hydrophilic head and a hydrophobic tail. In a hydrophilic environment, phospholipids spontaneously arrange into a double layer with the heads pointing toward the solvent and the tails packed against each other (Figure 8.1). The head group is different in different phospholipids and is attached to two fatty acid tails that can also differ in different membranes; in general, however, one tail does not contain double bonds and one does. This is important because the rigidity of a double bond keeps the two tails from tightly packing against each other and thus maintains the membrane sufficiently fluid at body temperature. Indeed this enables both lipid molecules and the embedded proteins to diffuse within the bilayer.

A serious bottleneck in prediction of the structure of membrane proteins is that the experimental determination of their structure is much more difficult than for soluble proteins and, as a result, the number of solved examples at our disposal is very limited. The membrane proteins constitute only approximately 1% of the known protein structures. This clearly makes the task of predicting their structure very difficult, because the size of the training set at our disposal is very poor. The membrane environment is, however, a constraint on the structure of these proteins and this can be exploited to develop methods for predicting their structure.

Protein Structure Prediction. Edited by Anna Tramontano
Copyright © 2006 WILEY-VCH Verlag GmbH & Co. KGaA, Weinheim
ISBN: 3-527-31167-X

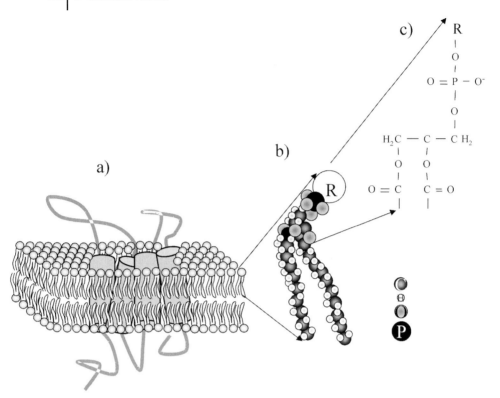

Figure 8.1 (a) Schematic diagram of a membrane with an embedded protein. (b) The structure of a phospholipid. (c) The chemical structure of its hydrophilic head. The phospholipid is formed by a glycerol molecule linked to two fatty acids, long linear carbon chains usually 16 or 18 carbon atoms in length, and a hydrophilic groups indicated by R in the figure. Glycerol is an alcohol with three carbon atoms, each bearing a hydroxyl group. Two of the hydroxyl groups of glycerol are bound to the fatty acids, which can be saturated or unsaturated, i.e. containing none or up to three double bonds, respectively. Two saturated hydrocarbon tails can pack tightly together An unsaturated fatty acid will have a kink in its shape wherever a double bond occurs; this results in looser packing and is important for the fluidity of the cell membrane. One of the hydroxyl groups of glycerol is joined to a phosphate group, which is negatively charged. Additional small molecules, usually charged or polar, can be linked to the phosphate group to form a variety of phospholipids. Because of the presence of a hydrophilic and hydrophobic region, phospholipids arrange in a double layer as shown in (a).

In solution, the backbone and the polar side-chains of a protein can hydrogen-bond with other amino acids or with solvent molecules. In the membrane interior, hydrogen-bonds can be formed only with the polypeptide chain itself. The energy cost of burying an unsatisfied hydrogen bond is very high, of the order of 4 kcal mol^{-1}, and this practically excludes the possibility of "random coil" structures within the bilayer. Secondary structures provide the means to saturate main-chain hydrogen-bond donors and acceptors – the main chain atoms of alpha helices are all involved in interchain hydrogen-bonds and so are the polar atoms of the main chain in a barrel of beta strands (Figure 8.2). Indeed all membrane proteins known so far are either alpha helical or form beta barrels, so there are no main chain polar atoms not involved in hydrogen-bonds. Membrane proteins of cells with a single membrane are all alpha helical; beta barrel membrane proteins are found only in the outer membranes of cells that have two membranes. As expected, the part of the polypeptide that faces the lipid comprises mainly hydrophobic side chains, because polar amino acids that, in solution, can form hydrogen-bonds with water are not favored in the membrane environment.

In the same way as for globular soluble proteins, prediction of secondary structure elements in membrane proteins can take advantage of methods such

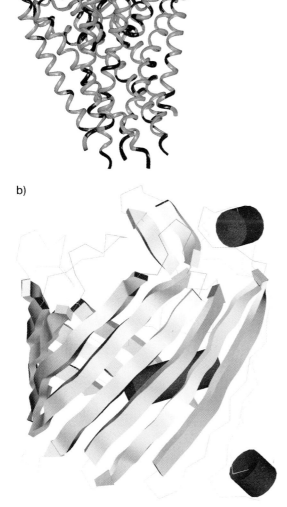

Figure 8.2 Two membrane proteins. The first is a potassium channel, the second a porin. They exemplify the two types of structure observed for membrane proteins – bundles of alpha helices, as in the channel, and beta barrels, as in the porin. In both examples main chain hydrogen-bonds are saturated.

as preference rules, neural networks, and support vector machines. Although the methods are the same, the number of available examples that can be used for training the algorithms is far smaller than for globular proteins. This is somewhat compensated by the higher regularity of these proteins dictated, as we have discussed, by the special environment in which they occur.

8.2
Prediction of the Secondary Structure of Membrane Proteins

An estimate of the likelihood that an amino acid is embedded in a lipid bilayer can be obtained by measuring partition coefficients. A partition coefficient is defined as the ratio of the equilibrium concentrations of a dissolved substance in the phases of a two-phase system consisting of two largely immiscible solvents, for example water and octanol. The ratio between the two concentrations is related to the free energy of transfer of the solute from one phase to the other. In practice, we can measure the equilibrium concentrations of an amino acid in a two-phase system consisting of water and an apolar solvent. If the two concentrations are $c1$ and $c2$, respectively, we know that:

$$c1/c2 = e^{-\Delta\mu/RT}$$

and the free energy of transfer $\Delta\mu$ can be computed as (Figure 8.3):

$$\Delta\mu = RT\ln(c2/c1)$$

This value reflects the preference of a free amino acid for a hydrophobic medium and can be used to estimate the propensity of each amino acid for the membrane environment.

The choice of the experimental system used for determining the partition coefficients affects the resulting hydrophobicity measure. The most relevant factors are the apolar solvent chosen and the chemical form of the amino acid used in the experiment. Historically, the first hydrophobicity scale was proposed by Nozaki and

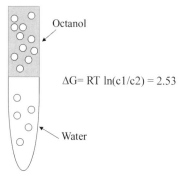

Figure 8.3 The free energy of transfer of a molecule between two media can be derived from the equilibrium concentrations of the molecule in a two phase system.

Tanford on the basis of the relative solubility of isolated amino acids in water and ethanol. The system is not ideal, because the carbon and nitrogen main-chain atoms of an isolated amino acid are in a different chemical form than in a protein chain (in which they are covalently bonded to other atoms) and ethanol is not a very apolar solvent. Several other experimental systems have since been used, for example partition coefficients have been measured for amino acids with the termini blocked with acetyl (CH$_3$C=O) and amide (NH$_2$) groups or for mimics of each amino acid with the Cα substituted with an hydrogen atom. None of these methods provides a very accurate estimate of the free energy of transfer of amino acids in a folded chain, because the local environment and, especially, the pattern of hydrogen bonds, substantially affects the results.

Other methods for estimating the hydrophobicity of amino acids are based on statistical analysis of their frequency in the internal core and on the surface of proteins of known structure, because it can be shown there is a linear correlation between the extent of surface area that a residue, or an atom, exposes to solvent, on average, and its hydrophobicity. As a result of these many attempts, more than one hydrophobicity scale have been reported in the literature and the debate about which is the most suitable for prediction of hydrophobic regions likely to be found in a membrane environment is ongoing.

Given a hydrophobicity s_i for each amino acid i, a protein sequence can be coded as a numerical sequence s_i, for $i = 1, L$ where L is the length of the protein. The sum of the s_i values for a contiguous sequence of N residues is an estimate of the average free energy (favorable or unfavorable) for burying the segment in the membrane:

$$H_N = \sum_{j=1}^{N} s_j$$

Dividing this value by N gives the average free energy for each individual residue in the segment:

$$H_{j,N} = \frac{1}{N} \sum_{j=1}^{N} s_j$$

One way to calculate the propensity along an entire amino acid sequence is by use of a sliding-window average. One computes the average hydrophobicity value for a window spanning residues 1 to N in the protein, then the window is moved one residue down the sequence and the average value for segment 2 to $N + 1$ is calculated, and so on until the end of the sequence is reached (Figure 8.4).

The value:

$$H_{i,N} = \frac{1}{N} \sum_{j=i}^{N+i-1} s_j \text{ for } i = 1, 2, 3, \ldots L-N+1$$

is assigned to the central residue of the window. Regions with high H values are likely to be embedded in the membrane. A typical bi-layer is 30 Å in width, each

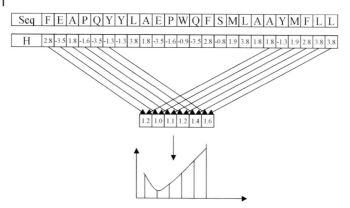

Figure 8.4 The method used to compute the hydrophobicity plot of a protein sequence. Each amino acid is assigned a hydrophobicity value. The average is computed over a sliding window of 19 residues.

turn of the helix spans ~5.4 Å, therefore at least 5.5 complete turns of helix (approx. 19 or 20 amino acids, because each turn of the helix contains 3.6 residues) are needed to traverse the membrane. Because of this, and because an odd number of residues is needed to define a central residue, N is usually set to 19. Figure 8.5 shows such a plot for a protein containing seven transmembrane helices.

These methods can, on average, predict more than 90% of known transmembrane helices, but only in 70% of instances are all the helices of a membrane protein correctly predicted.

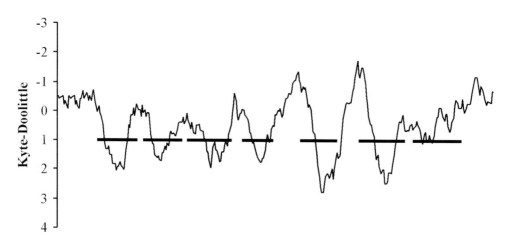

Figure 8.5 Hydrophobicity plot for the rhodopsin protein sequence. The black bars represent the experimentally determined position of the seven trans-membrane helices of this protein. Ideally, they should all correspond to large hydrophobicity values (minima in the plot).

8.3
The Hydrophobic Moment

The hydrophobicity plot only returns a scalar value for each residue, a different representation can be used to gain also spatial information, the helical wheel projection, i.e. a schematic representation of a helix with its axis perpendicular to the page, as shown in Figure 8.6.

We can calculate the hydrophobic moment, μ, of a helix as:

$$\mu_S = \sum_{j=1}^{N} s_j \vec{x_j}$$

where s_j is the hydrophobicity of the jth residue of the segment and x_j is a unit vector pointing from the alpha carbon of the amino acid to the center of its side-chain (Figure 8.6).

The hydrophobic moment is a vector sum of the directions of each side chain, weighted by its hydrophobicity. A hydrophobic residue contributes to the moment. A polar or charged residue reduces the moment if it points in the same direction in

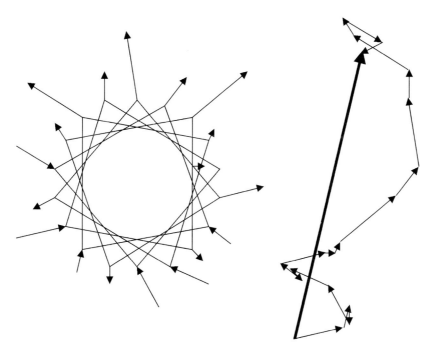

Figure 8.6 Helical wheel representation of a helix. The view is along the helix axis. The vectors represent the hydrophobic moment of each amino acid. The length of the vector is proportional to the hydrophobicity of the amino acid. The vectors connects the C alpha atom to the center of its side-chain. The hydrophobic moment of the helix is the sum of these vectors, indicated by the tick arrow on the left of the figure.

space as the hydrophobic residue, but increases it if it is in the opposite direction of the hydrophobic residue. If the structure of the protein under examination is not available, one can compute a hydrophobic moment for its sequence assuming it is folded as a alpha helix or a beta strand. In other words we can compute:

$$\mu_s = \left[\left[\sum_{j=1}^{N} s_j \sin(\delta n) \right]^2 + \left[\sum_{j=1}^{N} s_j \cos(\delta n) \right]^2 \right]^{1/2}$$

where δ, the angle at which successive side chains emerge from the central axis of the structure, is set to $\delta = 100°$ (for alpha helices) and $\delta = 180°$ (for beta strands) to test the hypotheses that the sequence is folded as an alpha helix or a beta strand. This enables us to establish whether the moment resulting from one of the two putative conformations is high.

Detection of beta barrels in membrane proteins is more difficult, because they usually have a central pore filled with polar solvent that can accommodate polar residues. Some methods use hydrophobicity plots only considering every other residue, but their accuracy is not very satisfactory.

8.4
Prediction of the Topology of Membrane Proteins

If we detect the location of transmembrane helices and strands, we can try to assemble them to derive the overall topology of the protein. This sequential strategy (first predict the secondary structure and next their three-dimensional arrangement) is not optimal, however. Prediction of the location of the secondary structure

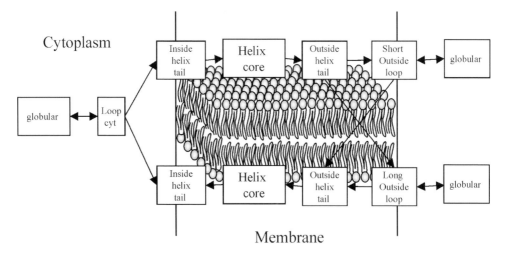

Figure 8.7 Possible scheme of a hidden Markov model for predicting the topology of a transmembrane protein. This is the layout used by the TMHMM server.

elements can be simplified by considering the overall topology of the protein, because other known properties of membrane proteins, for example the expected size of the connecting elements, can be taken into account. This strategy has been exploited by several automatic learning methods. For example, one can design a hidden Markov model with several states. For a helical transmembrane protein we can define three states (inside loop, helix, outside loop) and use proteins of known structure to train the model by estimating the transition and emission probabilities for each position along the sequence. Usually, these methods define five states ("inside loop", "inside helix tail", "helix", "outside helix tail", and "outside loop") to take into account the peculiar composition of regions at the border between the membrane and the polar environment. An example of the structure of such an HMM is shown in Figure 8.7.

The results of the prediction methods can be post-processed to improve their accuracy. For example, we know that loops are usually shorter than 60 residues and that inside loops are more positively charged than outside loops.

Neural networks can also be used effectively to predict the structure of membrane proteins. As already mentioned, these algorithms cannot be used to predict the three-dimensional coordinates of a protein because the relationship between the sequence and its coordinates depends on the specific frame of reference used and similar three-dimensional substructures in two proteins have a completely unrelated set of coordinates. The limited topology possible for membrane proteins implies we can attempt to predict the displacement of a residue along the axis of the alpha helices or of the beta barrel relative to the membrane boundary, and this can be sufficient to deduce the topology of the protein.

In the same way as for prediction of secondary structure, for prediction of membrane protein topology also evolutionary information in the form of multiple sequence alignments of homologous proteins can be used as input to the automatic learning algorithm and has beneficial effects on the accuracy of the methods. The accuracy of this type of method is approximately 80% for both helical and beta membrane proteins. The presence of membrane helices is correctly predicted for most proteins and most of the errors consist in predicting one helix too few or one too many. The results are usually more accurate for proteins with five or fewer membrane helices, and prediction accuracy drops for proteins with more than five helices.

More classical protein structure prediction techniques, for example fold-recognition and molecular dynamics, have been applied to membrane proteins, with promising results; it is, however, very difficult to estimate their performance accurately given the small number of structures available.

Suggested Reading

Three historical articles:

Y. Nozaki, C. Tanford (1971) Solubility of amino acids and 2 glycine peptides in aqueous ethanol and dioxane solutions – establishment of a hydrophobicity scale. J. Biol. Chem. **246**, 2211

D. Eisenberg, R. M. Weiss, T. C. Terwilliger (1984) The hydrophobic moment detects periodicity in protein hydrophobicity. Proc. Nat. Acad. Sci. **81**, 140–144

State of the art in membrane protein prediction:

B. Rost, R. Casadio, P. Fariselli, C. Sander (1995) Transmembrane helices predicted at 95% accuracy. Protein Sci. 4:521–533.

C.P. Chen, B. Rost (2002) State-of-the-art in membrane protein prediction. Applied Bioinformatics **1**, 21–35

B. Rost (2003) Prediction in 1D: secondary structure, membrane helices, and accessibility. Methods Biochem. Anal. **44**, 559–587

K. Mele, A. Krogh, G. von Heijne (2003) Reliability measures for membrane protein topology, J. Mol. Biol. **327**, 735–744

Some servers for predicting membrane proteins:

E.L.L. Sonnhammer, G. von Heijne, A. Krogh (1988) A hidden Markov model for predicting transmembrane helices in protein sequences. Proc. Sixth Int. Conf. on Intelligent Systems for Molecular Biology **1**, 175–182; http://www.cbs.dtu.dk/services/TMHMM/

M. Cserzo, E. Wallin, I. Simon, G. von Heijne, A. Elofsson (1997) Prediction of transmembrane alpha-helices in procariotic membrane proteins: the Dense Alignment Surface method. Prot. Eng. **10**, 673–676; http://www.sbc.su.se/~miklos/DAS/maindas.html

K. Hofmann, W. Stoffel (1993) TMbase – A database of membrane spanning proteins segments. Biol. Chem. Hoppe–Seyler **374**, 166; http://www.ch.embnet.org/software/TMPRED_form.html

G. von Heijne (1992) Membrane Protein Structure Prediction, Hydrophobicity Analysis and the Positive-inside Rule. J. Mol. Biol. **225**, 487–494; http://www.sbc.su.se/~erikw/toppred2/

I. Jacoboni, P. L. Martelli, P. Fariselli, V. De Pinto, R. Casadio (2001) Prediction of the transmembrane regions of beta barrel membrane proteins with a neural network based predictor. Protein Sci. **10**, 779–787; http://gpcr.biocomp.unibo.it/cgi/predictors/outer/pred_outercgi.cgi

B. Persson, P. Argos (1994) Prediction of transmembrane segments in proteins utilising multiple sequence alignments J. Mol. Biol. **237**, 182–192; http://www.mbb.ki.se/tmap/index.html

9
Applications and Examples

9.1
Introduction

Examples of instances in which protein structure modeling has provided invaluable information to experimentalists are countless and it is impossible to give a complete overview of all of them. The examples that will be described here were selected to give the reader a general overview, but unavoidably reflects the preference (and knowledge) of the author. Some examples have been selected for their historical importance, others for their medical relevance, yet others because they are instructive examples of effective collaboration between experimentalists and computational biologists. Many more, equally interesting, examples are described in the literature and together testify to the importance of structural models for the understanding of the biological properties of proteins.

9.2
Early Attempts

The first protein structure determined by X-ray crystallography was sperm whale myoglobin. This was achieved in 1958, through the work of Sir John Kendrew and coworkers (Figure 9.1). A few years later, in 1961, work started on experimental resolution of the structure of lysozyme. The work proceeded with measurement of intensities, correction of these for absorption, the preparation of heavy atom isomorphous derivatives, and use of anomalous scattering. The solution of the 2Å resolution structure of lysozyme was achieved in 1965 (Figure 9.2). The structure showed the complete polypeptide chain (129 amino acid residues), was folded into both alpha helices, previously seen in myoglobin, and beta sheet, a structure that had been predicted by Linus Pauling but not hitherto observed in three dimensions. The molecule was composed of two domains.

The structure of the lysozyme-tri-*N*-acetylchitotriose complex, which was published the same year, was very informative. It led to a detailed interpretation of the lysozyme-inhibitor complex and the key elements of recognition at the catalytic site. The next step was to work out how lysozyme recognized its substrate.

Protein Structure Prediction. Edited by Anna Tramontano
Copyright © 2006 WILEY-VCH Verlag GmbH & Co. KGaA, Weinheim
ISBN: 3-527-31167-X

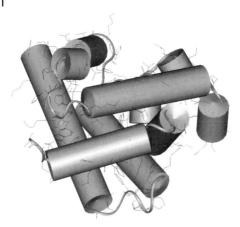

Figure 9.1 The structure of sperm whale myoglobin – the first protein structure determined by X-ray crystallography.

Molecular model building and available biochemical evidence were equally instrumental in enabling understanding of the way in which a hexasaccharide substrate must bind. For the first time, a structure provided an explanation of how an enzyme speeded up a chemical reaction in terms of physical chemical principles.

Even with only two protein structures available, comparative modeling gave its first, spectacular for the times, result. Browne and coworkers constructed a model for alpha-lactalbumin (Figure 9.3), on the basis of lysozyme, in 1969. At the time computer graphics was not readily available and the model was a physical one (made with balls and sticks). It was a rather good model and it was later established that the *rmsd* deviation of C alphas of the core between the model and the experimental structure was approximately 1 Å.

The principal steps in homology modeling were more rigorously established, in 1981, by Greer and colleagues who proposed the technique and applied it to a family of mammalian serine proteases (Figure 9.4).

These early attempts opened the road to the more sophisticated techniques described in this book, and established the principle that the crystal structure of a protein can provide information on the structure of many more proteins.

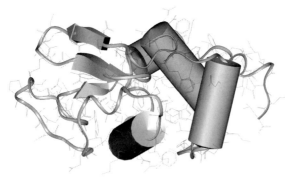

Figure 9.2 The structure of lysozyme – the first protein to be used as a template in a comparative modeling experiment.

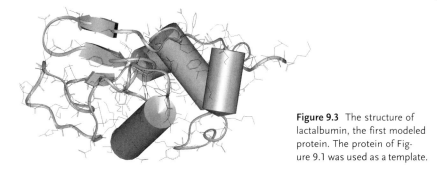

Figure 9.3 The structure of lactalbumin, the first modeled protein. The protein of Figure 9.1 was used as a template.

9.3
The HIV Protease

In 1978 the first cases of an infection followed by Kaposi sarcoma and Pneumocystis carcini was noted in the USA. The disease started killing patients in 1980 and was called gay related immunodeficiency disease, because its appearance seemed to be related to homosexual lifestyle. The disease quickly spread to the rest of the world and studies showed that it was a fatal disease resulting from viral infection. In 1983. Montagnier and Gallo discovered the connection between the HIV virus and the disease, named AIDS. There finally was a target. It was soon established that HIV interferes with the immune system by blocking a receptor on specialized cells called T helper cells.

HIV is a retrovirus. The genetic material of these viruses is RNA that is reverse-transcribed into DNA before transcription into mRNA, translation of which produces the viral proteins. The virus has an envelope (i.e. a lipid bilayer) and a core. The lipid bilayer contains the transmembrane protein gp41, the knob-like surface

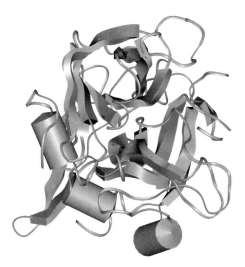

Figure 9.4 Blood coagulation factor Xa, a member of the mammalian serine proteases, modeled in 1981.

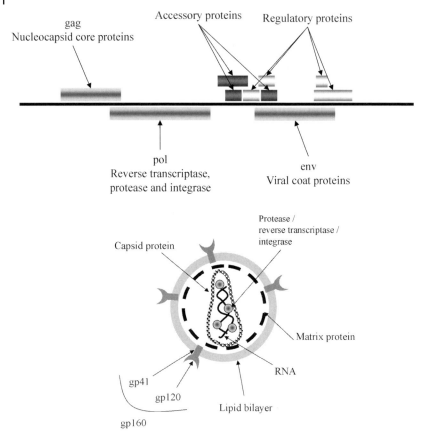

Figure 9.5 Genomic organization of the HIV-1 virus and schematic diagram of the virion.

glycoprotein gp120 on the outside, and the matrix protein p17 on the inside. The internal core of the virus is formed by the capsid protein p24 and contains two RNA molecules (Figure 9.5). The surface glycoprotein becomes attached to CD4 receptors on T helper cells. The transmembrane protein penetrates the cell and initiates membrane fusion and entrance. A virally encoded reverse transcriptase enzyme produces complementary single-strand DNAs. The DNA and a virally encoded integrase are transported to the nucleus and integrated into the host DNA (Figure 9.6).

In HIV the 5′ end of the *pol* gene codes for a protease used for processing of the *gag* polyprotein into the separate core proteins. Proteases can belong to different classes, as discussed in Chapter 1. Pepsin, for example, is an aspartic protease. It has two domains, and its active site, formed by two aspartic acids, is located in a deep cleft within the molecule.

In 1987 Pearl and Taylor examined the sequences of the retroviral protease, using pattern-recognition, structure prediction, and molecular modeling techni-

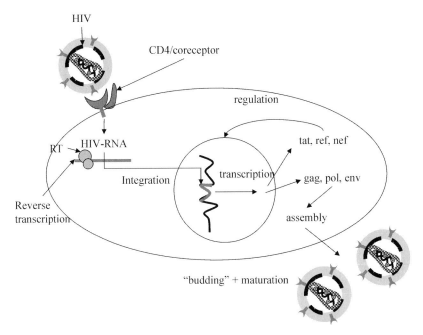

Figure 9.6 The life cycle of the HIV virus. Its proteins must be processed by the protease before assembly can occur.

ques, and formulated the unorthodox suggestion that it could fold as a single domain of an aspartic protease such as pepsin. Thus, the virus generates an active aspartic protease only when two identical 99-amino-acid domains join to create a

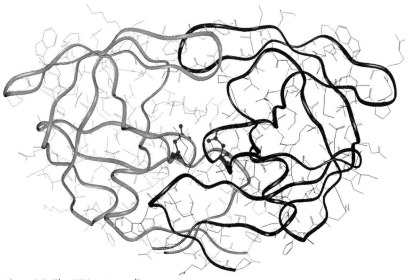

Figure 9.7 The HIV protease dimer.

homodimer (Figure 9.7). Determination of the HIV protease structure confirmed these predictions.

The first role of HIV protease is to cleave itself out of the large chain protein formed from the genetic material of the virus and, subsequently, to cleave out the remaining proteins of the virus. The function of HIV protease is therefore critical to viral replication and is an important target for drug development. Indeed AIDS therapy includes HIV protease inhibitors whose development has been accelerated by the understanding of the structure properties of the enzyme.

9.4
Leptin and Obesity

In 1953 it was proposed that the triacylglycerol content regulates body weight via alterations in food intake and energy expenditure through a centrally acting negative

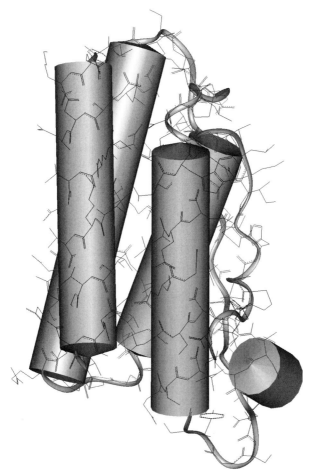

Figure 9.8 The structure of leptin. The three-dimensional model of the protein suggested it belongs to the class of helical cytokines.

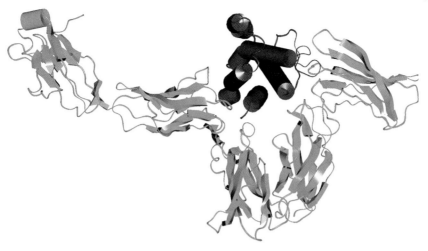

Figure 9.9 The complex between a helical cytokine (interleukin-6) and a portion of its dimeric receptor. The receptors of helical cytokines all have similar structural organization (immunoglobulin-like fold) and this led to the discovery of the leptin receptor.

feedback signal. At the end of the sixties, this hypothesis was substantiated by the discovery of the mutation obese (ob) in mice, which led to early onset of obesity. Later it was found that, if the blood systems of two mice were joined, overfeeding one led to reduced food intake by the other, supporting the hypothesis that a signaling molecule communicates how much fat is in the body to the brain's hypothalamus, which coordinates basic body functions such as eating. In 1994 the gene coding for the signaling protein was cloned. The protein was termed "leptin" from the Greek "leptos", meaning "thin". Sequence analysis revealed no similarities to other proteins and no clues about details of its functions. The remaining possibility was that *ob* is related to other proteins, whose sequences are divergent to the point that only a comparison of three-dimensional structures might detect relationship.

These were early times for fold recognition techniques, which can be dated to ca 1993–1994. Nevertheless Steve Bryant and colleagues attempted a "threading" search of a three-dimensional structure database to determine whether the *ob* protein might adopt a fold similar to any known structure. The search revealed that the *ob* sequence is compatible with structures from the family of helical cytokines. It was immediately recognized that such a relationship made biological sense. Cytokines are soluble proteins, released by cells of the immune system, which act non-enzymatically through specific receptors to regulate immune responses. They resemble hormones in that they act at low concentrations binding with high affinity to a specific receptor. The fold-recognition result not only led to a structural model of *ob*, later found to be correct (Figure 9.8), but also to the prediction that it could bind to a receptor resembling those of the cytokine family.

This was a very critical finding. Indeed, unlike the mice used to isolate the *ob* gene, obese humans do not have a mutated ob. Tests quickly showed that some obese people not only have enough leptin, but that they actually have levels 20 to 30 times higher than those found in lean people. This led researchers to hypothesize that overweight people might have a problem with the leptin receptor, rather than with leptin. The leptin might be telling the brain to stop eating, but the brain might not be able to listen if the receptor is not functional. The search for the leptin receptor started. It was finally found in 1995 (Figure 9.9). Scientists now had a target against which try to design drugs. The computational approach to structure prediction was undoubtedly a key player in this discovery process.

9.5
The Envelope Glycoprotein of the Hepatitis C Virus

More than 200 million people world-wide suffer from chronic hepatitis C, a very serious disease. Approximately 80% of patients develop chronic hepatitis, with 20% of these progressing to liver cirrhosis and hepatocellular carcinoma. The disease is caused by a virus, called HCV, whose genome contains a single open reading frame which encodes a polyprotein of approximately 3,000 amino acids. The structural proteins lie in the N-terminal portion of the polyprotein whereas the nonstructural proteins form the remainder (Figure 9.10). Polyprotein processing occurs via a combination of host and viral proteases and this gives at least ten unique proteins – C, E1, E2, p7, NS2, NS3, NS4A, NS4B, NS5A, and NS5B. C, E1 and E2 form the structural components of the virus whereas the other proteins are non-structural.

Similarly to all RNA viruses, HCV has a high mutation rate (approximately one nucleotide change per genome per generation) and exists as a quasi species. This is thought to be the mechanism by which the virus escapes the immune response and the reason for the difficulty encountered in developing an effective vaccine. HCV envelope glycoproteins are deemed to be very important, because chimpanzees immunized with purified recombinant E1/E2 proteins have been shown to be protected against challenge with homologous virus, i.e. with a virus containing the exact same E1/E2 sequence. No structural information is available on E2 and at the beginning of the nineteen-nineties no sequence similarity with any other protein could be detected. "Orphan" proteins, i.e. proteins for which there is no other related sequence available, are very difficult to analyze. As already discussed, most methods for sequence analysis, including those for predicting secondary structure, have very limited accuracy for single sequences.

Finally, in 1995, the sequence of two viruses related to HCV was obtained. The story of this discovery is interesting. In 1967, scientists inoculated tamarin monkeys with serum from GB, a surgeon suffering from acute hepatitis. Biochemical and histologic evidence of acute viral hepatitis in the monkeys became apparent. Both GB and the tamarins recovered from their hepatitis. After several passages, the genomes of two viruses were eventually cloned and sequenced in their entirety from the infectious tamarin serum These viruses were designated GB viruses A and B (GBV-A, GBV-B). Scientists searched the blood of patients with hepatitis for sequences similar to regions of GBV-A, GBV-B, and HCV and a positive individual was found in West Africa. The sequence of the virus infecting this individual was similar to, but distinct from, GBV-A and GBV-B and turned out to be a portion of the genome of a third virus, designated GB virus C (GBV-C) sequenced in 1996. In the same year, a new virus, very similar to GBV-C, designated HGV, was isolated from a patient diagnosed with non-ABC hepatitis. The GBV-C and HGV virus are now regarded as different isolates of the same virus.

Interestingly, GBV-A was later found in several species of New World monkeys. Most likely GBV-A and GBV-B did not originate from the initial GB inoculum and were probably present in tamarins before inoculation!

The discovery of three members of the HCV family was instrumental in pushing forward the discovery process. A combination of multiple sequence alignments, secondary structure prediction, and fold recognition methods applied to the E2 sequences of HCV, GBV-A, GBV-B, and GBV-C enabled detection of possible structural similarity between E2 and the envelope protein E of Tick Borne encephalitis virus (TBEV). The model was validated by mapping on to the model available experimental data on the location of known antibody epitopes and potential glycosylation sites. Manual adjustments of the fold-recognition alignments and careful inspection of the intermediate results of the model building procedure enabled a three-dimensional model of this protein to be produced. The model predicted the existence of a specific heparin-binding domain, subsequently experimentally verified, and enabled identification of putative interaction sites between E2 and its receptor. Subsequent experimental results indicated that at

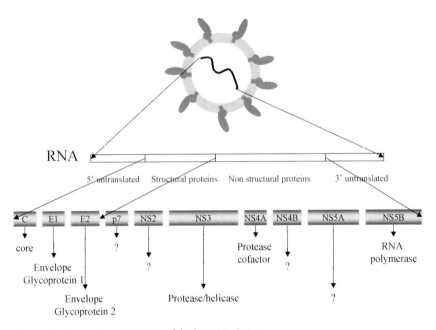

Figure 9.10 Genomic organization of the hepatitis C virus.

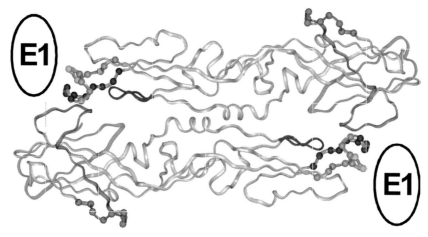

Figure 9.11 The model of the E2 protein of HCV. The colored regions correspond to parts of the protein known to be involved in binding with other molecules and, therefore, expected to be exposed to the solvent. The two ovals labeled E1 indicate where the model predicts the E1 protein binds.

least one of the proposed regions is, in fact, involved in the interaction. In addition, the authors noted that the TBEV E protein forms a dimer and therefore constructed a putative dimeric model of the HCV E2 protein. Analysis of this final model supported the hypothesis of the dimeric nature of E2, because it could explain other available experimental data, and also suggested a putative model for the HCV E1/E2 glycoprotein quaternary structure (Figure 9.11).

TBEV belongs to the *Flaviviridae* family, the same family to which HCV belongs. A corollary of the modeling results is, therefore, that these viruses are all related. This was later confirmed by determination of the crystal structure of the E2 protein of a member of a different family (alpha), Semliki Forest virus (SFV). Quite unexpectedly, the structure of the SFV E1 monomer closely resembles the TBEV E structure (Figure 9.12). These new data therefore retrospectively justifies the choice of TBEV E as a structural template for HCV E2.

Although the E2 model, because of the low sequence similarity with TBEV E, must be considered an approximation of the real structure, it provides a three-dimensional framework in which experimental data can be interpreted and tested and from which new hypotheses can be formed.

9.6
HCV Protease

The non-structural proteins of HCV are of major interest for the development of small molecule inhibitors. NS3 is known to encode for two enzymatic functions – a serine protease and an RNA dependent NTPase/helicase. Early analysis of multiple sequence alignments between the NS3 regions of HCV and other flaviviruses and pestiviruses predicted the N-terminal domain to contain a trypsin-like serine protease of approximately 180 aa, with a conserved His-Asp-Ser catalytic triad. The predictions were experimentally tested and proved the existence of a serine protease. Mutational analysis showed that the NS3 protease is responsible for cleavage at four sites within the protein, at the junctions between NS3/4A, NS4A/NS4B, NS4B/NS5A and NS5A/NS5B. The precise location of the cleavage and therefore the exact beginning and end of all the other HCV structural proteins was not known, however, which made it impossible to clone the proteins in order to study them *in vitro*.

Although the sequence similarity of the protease to any member of known structure of the serine protease family is extremely low, Pizzi and co-workers attempted to build an approximate homology model of its structure, by a combination of structural analysis of known protease structures, manual alignment, and careful model building. In practice, they constructed a multiple structural alignment of several known protease structures and used it to obtain a multiple sequence alignment to which the sequence of the viral protease was manually

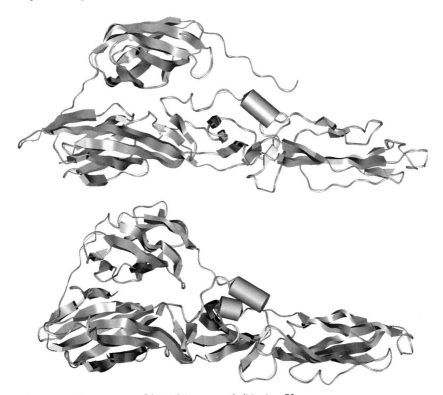

Figure 9.12 The structures of the Tick Born encephalitis virus E2 protein and of the Semliki Forest Virus E protein. Note that these proteins have a similar structure, although the two viruses are quite evolutionarily distant.

aligned. An important consequence of the procedure used is that the accuracy of the final model was different in different regions, depending on the extent of local structural similarity among the known proteases. This was taken into account when using the model to derive conclusions or plan experiments so that the authors were able to derive a wealth of very useful information from the model, notwithstanding its necessarily low overall accuracy.

As already mentioned, in comparative modeling the reliability of the predictions of functional regions is higher than that of the rest of the model, for evolutionary reasons, and indeed the NS3 protease model could be confidently used to analyze the specificity pocket of the enzyme, i.e. the part of the protein deputed to the recognition of the substrate. The analysis suggested the pocket was shallow, hydrophobic, and closed by the aromatic ring of a phenylalanine residue (Figure 9.13). A survey of pairs of interacting residues in the database of known protein structure highlighted that often the phenylalanine ring interacts with a cysteine, leading to the prediction that the substrates for NS3 protease should have a cysteine residue in the central position. The polyprotein of HCV did have at least one cysteine in the regions near the putative cleavage sites of the protease, supporting this conclusion; it was later experimentally confirmed by N-terminal sequencing the HCV proteins.

If a model is correct, one should be able to use it to predict the results of experiments. The reliability of this model was confirmed by altering the enzyme's specificity for a Cys to accept Phe instead.

Further multiple sequence alignment between the NS3 domains of HCV and of the GBV viruses highlighted three strictly conserved cysteine residues and one histidine in their sequences, which were mapped on to the model of HCV NS3. The relative positions of these residues in the model suggested they most probably formed a tetradentate metal-binding site. This was important in the light of earlier findings that mutation of the conserved cysteines compromises proteolytic activity of HCV NS3 and because it enabled a more efficient procedure for producing the enzyme to be established. Another important conclusion of the modeling project was that, despite a low (~30%) overall sequence identity with HCV NS3, the number of conserved residues in GBV-B NS3, including the catalytic triad and

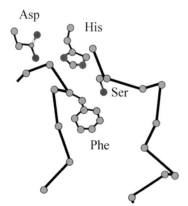

Figure 9.13 Schematic model of the specificity pocket of the NS3 protease of the HCV virus.

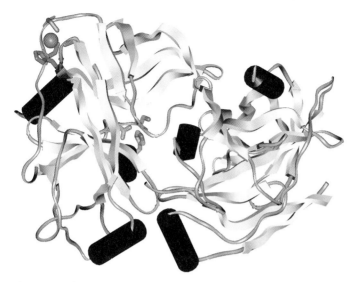

Figure 9.14 The X-ray structure of the NS3 protease of HCV. Notice the presence of the metal binding site predicted by the model.

the specificity pocket Phe, is sufficient to allow for shared protease substrate specificity for the two enzymes. Such information has been important in experiments successfully designing a chimeric HCV/GBV-B NS3 protein and in efforts to design a surrogate virus for testing of HCV inhibitors in animal models. Finally, X-ray crystallographic studies of the HCV NS3 protease showed that it adopted a chymotrypsin-like fold with a His-Asp-Ser catalytic triad, a specificity pocket containing a phenylalanine residue, and a structural zinc binding site involving three cysteine residues and one histidine thus fully confirming all modeling predictions (Figure 9.14).

The main lesson that can be learned from this example is that even very low accuracy models can be used effectively for advancing our understanding of a biological system, if their analysis takes into sufficient account the limitations of the procedure and the conclusions derived from the model are "weighted" according to the reliability of the region of the model on which they are based.

9.7
Cyclic Nucleotide Gated Channels

Ion channels are membrane-spanning proteins that allow ions, such as K^+, Na^+, Ca^{2+}, and Cl^-, to cross the hydrophobic cell membrane. The homotetrameric cyclic nucleotide-gated channel (CNG) from bovine rod, composed of the subunit CNGA1, forms functional assemblies involved in visual signal transduction. The protein can be activated by cyclic GMP which leads to an opening of the cation channel and thereby causes depolarization of rod photoreceptors. Each of the channel's subunits consists of two domains – a transmembrane domain formed by six transmembrane helices (S1–S6) and a pore helix (P-helix) (Figure 9.15) and a cytoplasmic domain formed by the cyclic nucleotide-binding domain linked to the transmembrane domain through a linker. The pore is believed to gate via a conformational change of the S6 transmembrane helix initiated by binding of cyclic nucleotides to the binding domains.

Giorgetti and coworkers combined experimental and computational approaches to provide a molecular basis for this proposal by constructing models of the transmembrane region of the channel. Models of P-helix-loop and S6 were based on the Kcsa X-ray structure, the topology of which had been suggested to be similar to that of CNG channels. The linker was modeled using the C-linker of HCN. This template shares a sequence identity (>30%) with CNG channels in this particular region. The model was based on a manually built alignment but was refined by the inclusion of many spatial constraints inferred from electrophysiological measurements; this makes the approach particularly interesting.

The idea was to construct mutants of the protein by substituting several of its residues with cysteines. Depending on their distance, pairs of cysteines can bind metal ions, or form disulfide bridges, or neither. Estimates obtained by calculation of residue–residue distribution in the Protein Data Bank suggested that the C alpha of cysteines located at approximately 11–13 Å can bind metals whereas two cysteines separated by a distance from 6 to 11 Å can form disulfide bridges. By measuring the metal-binding ability and disulfide formation for several mutants in their open and closed conformation, the authors derived constraints useful for building the final model and for formulating a functional model (Figure 9.16). They proposed that gating occurs by bending and rotation of the C-linker N-terminal section. This rotation is transmitted upwards, to the P helix which

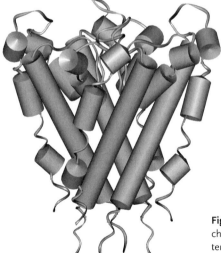

Figure 9.15 The structure of a potassium channel from *Streptomyces lividans* used as a template for the model of the nucleotide-gated channel CNG.

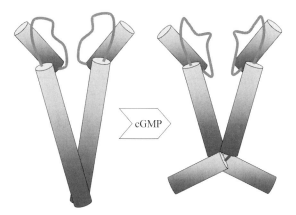

Figure 9.16 The model for the closed-to-open-state transition of CNG derived from a combination of computational and experimental analysis.

rearranges so that its terminal residues move away from the pore wall and lead to the opening of the pore lumen.

The conclusion from this work is that the initial event of cyclic nucleotide binding is transmitted to the pore walls by sophisticated coupling of conformational changes spanning the entire cytoplasmic and transmembrane domains of the channel.

This example, and those of the HCV proteins, demonstrate that a combination of experimental and computational approaches, and careful interpretation of the results, might lead to understanding of very complex biological mechanisms.

9.8
The Effectiveness of Models of Proteins in Drug Discovery

Every large pharmaceutical company has a huge collection of compounds, synthesized for different projects, bought from chemistry laboratories, collected as reaction intermediates in synthetic processes, or derived from natural sources. High-throughput screening of these chemical libraries is the most common method for finding new lead compounds in drug discovery. The time and effort needed for screening are substantial, however. A possible solution is to computationally select a smaller subset of compounds likely be enriched in compounds able to bind to the target. This preselection can be done by virtual screening, a computational method for selecting the most promising compounds from an electronic database for experimental screening (Figure 9.17). Virtual screening can be performed by searching databases for molecules either fitting a pattern derived from the analysis of a set of active compounds (pharmacophore) or by docking the compounds to a three-dimensional structure of the macromolecular target.

Pharmacophore-based screening has been extensively used and has the advantage of being applicable to ligands for which the target three-dimensional structure

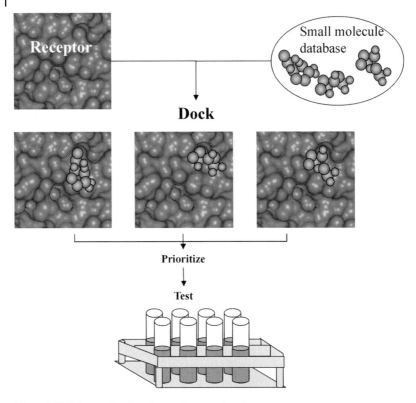

Figure 9.17 Scheme of a virtual screening experiment.

is unknown, but it requires the availability of a diverse set of compounds known to bind the target. Structure-based virtual screening, on the other hand, suffers from the limited accuracy of our docking methods and requires knowledge of the structure of the target. It will be very important in the near future to be able to use not only X-ray or NMR structures, but also protein models for protein-based virtual screening of chemical libraries.

There are already examples of successful use of proteins in the design of non-peptidic small inhibitors. These include several proteases, for example those from HCV, malaria, schistosoma, and cancer related proteins, etc. Another class of molecules that has been intensely investigated is the protein kinases. The catalytic domain of kinases is very well conserved structurally, and it is therefore possible to construct comparative models for them. The challenge here consists in developing specific inhibitors, because the high level of sequence conservation increases the possibility that drugs developed against one kinase also affect other, unrelated proteins of the same class. A few success stories have been reported in the literature, for example the discovery of an inhibitor of human CDK4, a member of the important cyclic-dependent kinase family that plays a central role in regulation of the cell cycle.

9.8 The Effectiveness of Models of Proteins in Drug Discovery

G-protein-coupled receptors are a special case. These are seven transmembrane-spanning receptors that couple to heterotrimeric G proteins (short for guanine nucleotide binding proteins) and are by far the largest receptor superfamily in our genome, mediating functions across the spectrum of physiology. Because of this, these membrane proteins plays an enormously important role in drug development – approximately 40% of all drugs are directed to a member of this class. Because they are very difficult to crystallize, determination of the crystal structure of one G-protein-coupled receptor (rhodopsin) in 2000 was a landmark in the study of GPCR, opening the road to the possibility of using comparative modeling to predict the structure of many more proteins of the family (Figure 9.18).

Unfortunately, many GPCR of interest for drug development share rather low sequence identity with rhodopsin. Whereas the transmembrane helices can frequently be predicted with reasonable certainty, the extracellular loops are much more divergent in sequence and very difficult to model. Extensive testing of docking methods with rhodopsin-based comparative models, using different docking programs and different scoring functions are performed continuously. It seems that use of a "consensus" scoring procedure for the predicted protein-ligand complexes involving several scoring functions provides the best results and produces encouraging results.

Our ability to use models of these proteins as targets in virtual screening is, however, still made very difficult by three of the problems we have repeatedly discussed in this book – the small number of available structures of membrane proteins, our limited ability to predict non-regular elements of protein structure, and the inadequacy of our potential energy functions.

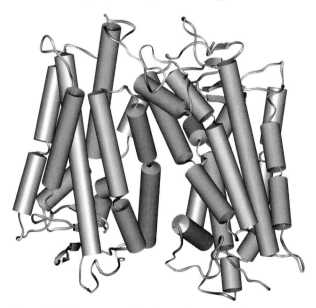

Figure 9.18 The structure of rhodopsin, the template for many modeling experiments of G protein-coupled receptors.

In summary, for virtual screening good overall accuracy of a model is not sufficient, we need to predict the precise positioning of side-chains in the binding site. Here the devil is in the detail and, as already mentioned, attempts to refine models relatively close to the native structure have met with little success. As we will see in the next section, virtual screening is not the only application of modeling in which improving the accuracy of a model, even marginally, could make a difference.

9.9
The Effectiveness of Models of Proteins in X-ray Structure Solution

We discussed X-ray crystallography in Chapter 1. Let us recall here that, in a diffraction experiment, a crystal is irradiated with a particular X-ray wavelength and the resulting diffracted waves are collected on physical or electronic devices. In this passage from 3D to 2D, however, all information about the phase of the diffracted waves is lost, which is one of the fundamental problems of structural science. We have mentioned that three approaches are used to solve the phase problem – direct methods, interference-based methods, and molecular-replacement methods. The last method is based on the fact that prior knowledge of the structure of a protein simplifies the solution of a different crystal form of the same molecule. Sometimes, the structure of a homologous protein or a model of the target protein can be sufficient to approximate the relative position of the atoms in the structure and enable the phases to be computed.

Historically, it has been very difficult to decide a priori the quality of a model that is required for a successful molecular replacement experiment. A generally accepted "rule of thumb" is that molecular replacement is effective if the model is reasonably complete and shares at least 40–50% sequence identity with its template structure. A recent computational experiment has shown that matters are more complex than this, however. The experiment consisted in using a set of models deposited in the CASP database as input for a completely automatic molecular replacement procedure and recording when the model was sufficient to obtain the phases and, therefore, to solve the structure. The conclusions of the work can be summarized as follows:

- In this specific application, what really counts is the overall quality of the model rather than the details of the less well predicted parts.
- A GDT-TS (see Chapter 2) greater than 84 is always sufficient to guarantee the success of the procedure, irrespective of the sequence identity between the target and template structure, of the method used for producing the model and of the structural class of the protein under examination. In the automatic procedure, models with GDT-TS below 80 were never successful.
- For models of intermediate quality, the results vary. A large fraction of the structure can usually be automatically built with respectable quality and it is likely that, in these cases, more iterations and, most of all, manual intervention can lead to success. Even limited improvements in the quality of a model can be

instrumental in the success of a molecular replacement experiment. This observation can explain why it has been so difficult so far to predict beforehand when a model can be successfully used in molecular replacement solely on the basis of the sequence identity between a model and the structural template used to build it.

This is yet another example in which differences between model quality can make a substantial difference to the outcome; it implies that more effort should be devoted to improving the initial model, because even minor improvements can be important. This aspect has been recognized by the CASP community who agreed that in the future more importance should be given to the details of the models produced.

The CASP experiment is not only a useful tool for evaluating methods, but it also provides scientists with a large set of data collected over the years as a result of the efforts of hundreds of experts. This is an invaluable tool that can help the prediction community to understand the limits of current modeling methods and emphasize the areas where effort can be more productively focused.

Suggested Reading

The first applications of modeling reported in the literature:

W.J. Browne, A. C. North, D. C. Phillips, K. Brew, T. C. Vanaman, R. L. Hill (**1969**) A possible three-dimensional structure of bovine alpha-lactalbumin based on that of hen's egg-white lysozyme. J. Mol. Biol. **42**, 65–86

J. Greer (**1981**) Comparative model-building of the mammalian serine proteases, J. Mol. Biol. **153**, 1027–1031

The first three-dimensional structures of proteins:

J.C. Kendrew, G. Bodo, H. M. Dintzis, R. G. Parrish, H. Wyckoff, D. C. Phillips (**1958**) A three-dimensional model of the myoglobin molecule obtained by x-ray analysis. Nature **181**, 662–666

H. Muirhead, M. F. Perutz (**1963**) Structure of haemoglobin. A three-dimensional Fourier synthesis of reduced human haemoglobin at 5.5 Å resolution. Nature **199**, 633–638

C.C. Blake, D. F. Koenig, G. A. Mair, A. C. North, D. C. Phillips, V. R. Sarma (**1965**) Structure of hen egg-white lysozyme. A three-dimensional Fourier synthesis at 2 Ångstrom resolution. Nature **206**, 757–761

The examples described here are reported in the following articles:

L.H. Pearl, W. R. Taylor (**1987**) A structural model for the retroviral proteases. Nature **329**, 351–354

T. Madej, M. S. Boguski, S. H. Bryant (**1995**) Threading analysis suggests that the obese gene product may be a helical cytokine. FEBS Lett. **373**, 13–18

A.T. Yagnik, A. Lahm, A. Meola, R. M. Roccasecca, B. B. Ercole, A. Nicosia, A. Tramontano (**2000**) A model for the hepatitis C virus envelope glycoprotein E2. Proteins **40**, 355–366

E. Pizzi, A. Tramontano, L. Tomei, N. La Monica, C. Failla, M. Sardana, T. Wood, R. De Francesco (**1994**) Molecular model of the specificity pocket of the hepatitis C virus protease: implications for substrate recognition. Proc. Natl. Acad. Sci. **91**, 888–892

C.M. Failla, E. Pizzi, R. De Francesco, A. Tramontano (**1996**) Redesigning the substrate specificity of the hepatitis C virus NS3 protease. Fold Des.**1**, 35–42

A. Giorgetti, P. Carloni (**2003**) Molecular modeling of ion channels: structural predictions. Curr. Opin. Chem. Biol. **7**, 150–156

A. Giorgetti, D. Raimondo, A. E. Miele, A. Tramontano (**2005**) Evaluating the usefulness of protein structure models for molecular replacement Bioinformatics, 21, ii72–ii76.

Conclusions

A knowledge, even approximate, of the three-dimensional structure of a protein is essential for understanding the details of its molecular function and gives valuable insights for development of effective rational strategies for experiments such as studies of disease-related mutations, site-directed mutagenesis, or structure-based drug design. The number of known protein structures is an order of magnitude smaller than the number of protein sequences that can be deduced from genome data. The only way to bridge this gap is to resort to computational methods such as those described in this book. These methods have matured, and information from three-dimensional protein models has been successfully used in a wide variety of biomedical applications, as we have discussed. There is no doubt that structure prediction methods are an essential part of the scientific background and toolbox of life scientists. The need for integrating experimental knowledge with theoretical hypotheses will only grow in the future: it has been estimated that every new experimental structure carries information about the structure of at least a hundred other proteins.

The hope is that this book has convinced the reader that computational methods are essential tools for research, rather than annoying chores that interrupt the experimental work, or ways of having a relaxing break, and that it is important to grasp their advantages and limitations, even though their use only requires a few "clicks". This book should have conveyed three main messages to its readers.

- First, structure prediction methods are continuously being developed and evaluated and it is important to keep up with the latest developments. This can make a difference to a scientific project, greatly accelerate the discovery process, and provide a structural framework for new hypotheses.
- Second, different methods have different reliability and it is possible to evaluate a priori whether a method can provide a model sufficiently accurate for its foreseeable use. The expected use of a model dictates how accurate it needs to be. This, in turn, defines whether the available modeling methods are suitable for the project.
- Third, the reliability of a protein structural model is not uniform. Some regions can be predicted better than others, and consequently can be used to derive information at different level of detail. This must be taken into account, and is why it is important to understand the details of structure prediction methods.

In summary, here are a few questions that should be asked before starting a modeling exercise:
- What do I need the model for?
- What level of accuracy does my problem require?
- Does the expected accuracy of the model match that required?
- Which experiments will I perform to validate the model?

Finally, a word of caution is necessary. A protein model is not an experimentally determined structure. Every time you look at the image of a protein model, no matter how beautiful and convincing it looks, you should keep in mind Magritte's famous painting showing a pipe and the hand-written sentence *"Ceci n'est pas une pipe"*. And indeed nobody can smoke it. Even if a model has been built using very reliable techniques, it must be validated experimentally and indeed, the most useful models are those originating from a synergy between computational and experimental efforts, establishing a virtuous circle in which the model suggests experiments the results of which are used to validate and refine the model.

Glossary

Acid: A substance which ionizes in aqueous solution to yield H^+. It can also be defined as an electron-pair (or proton) donor.

Analogy: Two elements that resemble each other, although they do not share a common evolutionary ancestor, are called analogous. (See also "homology")

Ångström (abbreviated Å): Unit of length. 10^{-10} m = 0.1 nm. It is not part of the International System of Units (the meter-kilogram-second system), but its usage is generally accepted in protein science.

Antibody (also called "immunoglobulin"): Protein secreted by the immune system in response to infection. Antibodies are produced by B cells. These cells mature in the bone marrow and express a specific membrane-bound antibody. When the membrane-bound antibodies interact with an antigen, B-cells differentiate into antibody-secreting plasma cells and memory cells. (See also "antigen" and "immune response")

Antigen: A substance that is recognized by the immune system as foreign to the body.

Base: A substance which ionizes in aqueous solution to yield OH^-. It can also be defined as an electron-pair (or proton) acceptor.

Biological membrane: A biological membrane or biomembrane is a membrane which acts as a barrier around a cell or one of its compartments. It is composed of a double layer of phospholipids (q.v.). Membranes contain proteins that can act as channels or receptors for external stimuli.

Catalytic site: The region of an enzyme where a chemical reaction occurs. An enzymes accelerates a reaction by reducing the activation energy of the reaction, i.e. by stabilizing high-energy intermediates. The active site is the set of residues that provide the stabilization.

Cell cycle: The sequence of stages that a cell passes through between one cell division and the next. The cell cycle is divided into M, G, S, and G2 phases. In the M phase, nuclear and cytoplasmic division occur. In the G phase there is a high rate of biosynthesis and growth; in the S phase duplication of the DNA content occurs as a consequence of chromosome replication; in the G2 phase the final preparations for cell division are made.

Chloroplast: The organelle in a plant cell that contains chlorophyll and is the site of photosynthesis.

Chromosome: A structure of compact, intertwined molecules of DNA (deoxyribonucleic acid) found in the nucleus of cells which carry the cell's genetic information. Humans normally have 46 chromosomes.

Circular dichroism spectroscopy: Circular dichroism, or CD, is the differential absorption of left and right-handed circularly polarized light by optically active molecules such as some sugars and amino acids. Because secondary structure (α helices, β strands, the double helix of nucleic acids) also has an effect on the CD of a molecule in the ultraviolet region (190–250 nm), the ultraviolet CD spectra of proteins can be used to measure their secondary structure content and, also, to measure changes in their conformation. It can, for instance, be used to study how the secondary structure of a molecule changes as a function of temperature or of the concentration of denaturing agents. Circular dichroism in the near-UV spectrum (250–350 nm) is used to monitor the tertiary structure of proteins. At these wavelengths, the chromophores (i.e. the groups of atoms that selectively absorb the light) are the aromatic amino acids and the disulfide bonds, and the CD signals they produce are sensitive to the overall tertiary structure of the protein.

Clinical trial: The systematic investigation of the effects of specific treatments according to a research plan in patients with a particular disease or class of diseases. Clinical trials have several phases. Phase I involves healthy volunteers and is used to verify whether the metabolic and safety data obtained in laboratory testing can be extrapolated to humans. The next phase, phase II, involves a few hundred patients and is aimed at establishing the dosage regimen and at studying the pharmacokinetics of the drug. In Phase III, the treatment is tested on a few thousand patients and its efficacy evaluated and compared with other available treatments. In the last phase, the efficacy and safety of the drug are monitored after its commercial introduction.

Core of a protein: The region of a protein which is more conserved during evolution. It usually corresponds to internal well packed regions and to main secondary structural elements.

Cytokines: Protein molecules that regulate interactions in the immune system. They are messengers that carry biochemical signals to regulate local and systemic immune responses, inflammatory reactions, wound healing, formation of blood cells, and many other biological processes.

Dielectric constant: see "Permittivity"

Distance: A distance is a function that satisfies specific rules. The distance between two elements A and B is always positive if A and B are different and zero if they are the same, and is independent on the order (i.e. the distance between A and B is the same as the distance between B and A). If the function also satisfies the inequality distance (A,B) < distance (A,C) + distance (C,A) it is called a metric distance. We are all familiar with the Euclidean distance defined as:

$$\text{distance}(A, B) = \sqrt{(x_A - x_B)^2 + (y_A - y_B)^2 + (z_A - z_B)^2}$$

where (x_A, y_A, z_A) and (x_B, y_B, z_B) are the Cartesian coordinates of the two points A and B, respectively. Euclidean distance is metric and indeed, in a triangle, one side is always shorter than the sum of the other two.

Eigenvalue: A scalar value λ is called the eigenvalue of a square matrix **A** is if there exists a vector **X** such that **AX** = λ**X**. (See also "eigenvector" and "matrix operations")

Eigenvector: A vector **X** is called the eigenvector of a square matrix **A** if there exists a scalar λ such that **AX** = λ**X**. (See also "eigenvalue" and "matrix operations")

Electric dipole: An electric dipole is a system comprising two charges of equal and opposite sign separated by a distance. The electric dipole moment for such a pair of opposite charges of magnitude q is the magnitude of the charge multiplied by the distance between them; the defined direction is toward the positive charge. If a material contains polar molecules, they will generally be in random orientations when no electric field is applied. An applied electric field will polarize the material by orienting the dipole moments of polar molecules. This increases the permittivity or dielectric constant of the material (q.v.) and therefore reduces the effective electric field.

Enthalpy: A thermodynamic quantity (with units of energy), symbol H, and defined by $H = E + PV$, where E is the internal energy of the system (i.e. the energy related to the motion and configuration of its atoms, molecules, and subatomic particles), P its pressure, and V its volume. At constant pressure, which is usually true for biological systems, enthalpy measures the quantity of heat that flows in or out of a system, hence differences in enthalpy can be measured directly by calorimetry.

Entropy: In statistical mechanics entropy is a measure of the "disorder" of a system, i.e. the number of available configurations or microscopic states that are consistent with a given macroscopic or average state. The entropy, S, of a system is given by the formula inscribed on Boltzmann's tombstone: $S = k \ln \omega$, where k is a constant and ω the number of configurations of the system.

Extreme value distribution: A family of statistical distributions (q.v.) with the same general form:

$$p(x) = 1 \bigg/ \beta e^{\frac{x-\mu}{\beta}} e^{-e^{\frac{x-\mu}{\beta}}}$$

where β is positive and μ is the location parameter (related to the "position of the distribution" with respect to the x axis).

Eukaryote: Cell or organism with membrane-enclosed, structurally distinct nucleus. Eukaryotes include all organisms except viruses, bacteria, and blue-green algae. (See also "prokaryote")

Evolution: The change in the genetic material of a population of organisms over time. This process is driven by mutations and directed by natural selection, i.e. by the differential survival and reproduction of organisms that enable them to adapt to their environment.

Exon: Segment of an interrupted gene that is represented in the mature RNA product. (See also "intron")

Fibrinopeptide: Peptide released from the amino end of fibrinogen by the action of thrombin to form fibrin during clotting of the blood. The removal of the peptide causes fibrin to assemble into large protein networks which trap blood cells, blocking the damage.

Free energy: Free energy is a measure of the capacity of a system to do work, such that a reduction in free energy could in principle yield an equivalent quantity of work. It is defined as $G = H - TS$ where H, T, and S are the enthalpy, absolute temperature, and entropy (q.v.) of the system.

Gaussian distribution (Also called normal distribution): A family of statistical distributions (q.v.) with the same general form:

$$p(x) = \frac{1}{\sigma\sqrt{2\pi}} \exp\left(-\frac{(x-\mu)^2}{2\sigma^2}\right)$$

where μ is the mean (average of the values of the distributions) and σ the standard deviation of the data defined as:

$$\sigma = \sqrt{\frac{1}{n}\sum_{i=1}^{n}(x_i - \mu)^2}$$

where x_i are the observed values and μ their mean.

The standard normal distribution is the normal distribution with a mean of zero and a standard deviation of unity.

Gene: A unit of nucleic acid that carries information for the biosynthesis of a specific product in the cell. Genes are contained by, and arranged along, the chromosome.

Glycoprotein: A molecule that consists of a carbohydrate (sugar) plus a protein. Glycoproteins play essential roles in the body. Almost all the key molecules of the immune system (q.v.) are glycoproteins.

G-protein coupled receptors (abbreviated GPCRs): Integral membrane proteins formed by seven transmembrane helices. GPCRs are found in a very wide range of species, and are always involved in signaling from outside the cell to inside the cell. GPCRs detect a signal (for example the binding of a small molecule, a peptide, or a photon) and transmit it to the cytosolic side causing activation of a G protein (a protein that binds GTP).

Hemoglobin: A protein found in the red blood cells that is responsible for carrying oxygen around the body. Hemoglobin picks up the oxygen in the lungs and then releases it in the muscles and other tissues where it is needed.

Heuristic: In computer science, a heuristic is an algorithm for which there is no formal proof it finds the correct solution, but that usually finds reasonably good approximate solutions in reasonably short time. In other words, is an algorithm intended to gain computational performance or conceptual simplicity, potentially at the cost of accuracy or precision.

Homology: Two elements (genes, proteins, anatomical structures) are called homologous if they are derived from a common evolutionary ancestor. (See also "analogy", "orthology", and "paralogy")

Hydrolysis: Addition of the elements of water to a substance, often with breakdown of the substance into two parts, for example the hydrolysis of an ester to an acid and an alcohol.

Hydrophilicity: The property of polar molecules to form water's hydrogen-bonded structures and, therefore, to be relatively water-soluble. (See also "hydrophobicity")

Hydrophobicity: The property of nonpolar molecules to disrupt the hydrogen-bonded structure of water without forming favorable interactions with the water molecules. These molecules are insoluble in water. (See also "hydrophilicity")

Immune response: The general reaction of the body to substances that are foreign or treated as foreign. (See also "antibody", "antigen")

Immunoglobulin: see "Antibody"

Immunoglobulin fold: Protein architecture characterized by a sandwich of two β sheets each usually formed by seven strands, although the number of strands can differ slightly in different families.

Intron: The nucleic acid sequence interrupting the protein-coding sequences of a gene (q.v.); intron sequences are transcribed into RNA but are cut out of the message before it is translated into protein. (See also "exon")

Magnetic field: Magnetic fields are produced by electric currents. They are usually described in terms of their effect on electric charges. A moving electric charge will accelerate in the presence of a magnetic field, changing its velocity and direction of travel because it will experience a force, $\vec{F}$, given by the Lorentz equation:

$$\vec{F} = q\vec{v} \times \vec{B}$$

where $\vec{v}$ is the velocity of a particle of charge q and $\vec{B}$ the magnetic field.

The equation implies that the force is perpendicular to both the velocity of the charge q and to the magnetic field $\vec{B}$, so the charged particle will be pushed in a direction perpendicular to the magnetic field and the direction of motion. It also implies that the magnetic force on a stationary charge or a charge moving parallel to the magnetic field is zero.

Matrix: A matrix is a tabular representation of a set of data. It is characterized by its dimensionality measured by the number of rows (m) and columns (n). If $m = n$ the matrix is said to be square, and if the upper triangle of values is identical with the lower triangle of values it is said to be symmetric.

Matrix operations: The most common operations involving matrices (q.v.) are transposition, sum, difference, scalar multiplication, matrix product and inversion. Given a matrix **A** with m rows and n columns, usually described as an $m \times n$ matrix, we call the transpose of **A**, denoted $\mathbf{A}^T$ or $\mathbf{A}'$, a matrix in which the rows and columns are inverted, or transposed. For example, the transpose of the matrix

$$\begin{pmatrix} 2 & 3 & 0 \\ 1 & -1 & 6 \end{pmatrix} \text{ is } \begin{pmatrix} 2 & 1 \\ 3 & -1 \\ 0 & 6 \end{pmatrix}.$$

The sum of two matrices **A** and **B** with the same number of rows and columns is a matrix in which each element is the sum of the corresponding elements in the two matrices, i.e.

$$\begin{pmatrix} 2 & 1 \\ 3 & -1 \\ 0 & 6 \end{pmatrix} + \begin{pmatrix} 1 & 0 \\ 4 & 3 \\ -2 & 1 \end{pmatrix} = \begin{pmatrix} 3 & 1 \\ 7 & 2 \\ -2 & 7 \end{pmatrix}.$$

Matrix subtraction works the same way:

$$\begin{pmatrix} 2 & 1 \\ 3 & -1 \\ 0 & 6 \end{pmatrix} - \begin{pmatrix} 1 & 0 \\ 4 & 3 \\ -2 & 1 \end{pmatrix} = \begin{pmatrix} 1 & 1 \\ -1 & -4 \\ 2 & 5 \end{pmatrix}.$$

We can multiply a matrix by a scalar simply by multiplying each of its elements by the scalar value:

$$3 \times \begin{pmatrix} 2 & 1 \\ 3 & -1 \\ 0 & 6 \end{pmatrix} = \begin{pmatrix} 6 & 3 \\ 9 & -3 \\ 0 & 18 \end{pmatrix}.$$

Multiplication of two matrices can be performed only if the number of columns of one matrix is equal to the number of rows of the other, because the value of the element (i,j) of the product matrix is given by the sum of the products of the corresponding entries in rows and columns. In practice, one first multiplies the first element of the row j with the first element of the column j, then the second element of the row j with the second element of the column j, etc., and finally adds all the numbers:

$$\begin{pmatrix} 1 & 6 \\ 9 & 3 \end{pmatrix} \times \begin{pmatrix} 0 & -1 \\ -1 & 2 \end{pmatrix} = \begin{pmatrix} -6 & 11 \\ -3 & -3 \end{pmatrix}.$$

The element in the first row and first column of the product is given by:

$$(1) \times (0) + (6) \times (-1) = -6.$$

The element in the first row and second column of the product is given by:

$$(1) \times (-1) + (6) \times (2) = 11$$

Messenger RNA (mRNA): The RNA that encodes and carries information from DNA to sites of protein synthesis. RNA polymerase makes a copy of a gene from the DNA to mRNA. Although this process, called transcription, is similar in eukaryotes (q.v.) and prokaryotes (q.v.), subsequent processing is very different in the two groups. Prokaryotic mRNA is essentially ready upon transcription and, usually, requires no further processing. In eukaryotes, a modified guanine nucleotide is added to the "front" of the pre-mRNA (i.e. to the mRNA as it comes out of transcription). This modification is needed for attachment of mRNA to the ribosome. Next, pre-mRNA is modified to remove stretches of non-coding sequences called introns (q.v.); the stretches that remain include protein-coding sequences and are called exons (q.v.). Sometimes pre-mRNA messages may be spliced in several different ways, enabling a single gene to encode multiple proteins. This process is called alternative splicing. In eukaryotes and sometimes also in prokaryotes, a stretch of adenines is linked to the messenger RNA molecule at its 3' end. Occasionally, an mRNA will be edited, changing the nucleotide composition of that mRNA. An example in humans is apolipoprotein B mRNA, which is edited in some tissues, but not others. In this protein, the editing creates an early stop codon which, on translation, produces a shorter protein.

In eukaryotes, in which there is a nucleus, mRNAs must also be exported from the nucleus to the cytoplasm before being translated in their products by the ribosome (q.v.).

Metabolism: The sum of all the physical and chemical processes by which living organized substance is produced and maintained (anabolism) and by which energy is made available to the organism (catabolism).

Mitochondria: Membrane-enclosed cellular compartments that are the major source of a cell's energy. Mitochondria contain some independent DNA genes which are maternally inherited.

Myoglobin: The molecule that accepts oxygen and stores it temporarily in muscle fibers until it is needed for metabolism there.

Nitrogenous bases: Organic compounds that owe their basic properties to the lone pair of electrons of a nitrogen atom. Typical nitrogenous bases are ammonia (NH_3), triethylamine, pyridine, and the nucleic acid bases adenine, guanine, thymine, and cytosine.

Nuclear spin: Intrinsic property of some nuclei that gives them an associated characteristic angular momentum and magnetic moment. Nuclear angular momentum is quantized (i.e. can take certain values only) as integral or half-integral multiples of $(h/2\pi)$, where h is Planck's constant (6.626×10^{-34} J s).

Orthology: An evolutionary relationship between two genes that only involves speciation and does not involve duplications. (See "homology")

Paralogy: An evolutionary relationship between two genes that involves duplications (See "homology")

Permittivity (or dielectric constant): A quantity that describes how an electric field affects and is affected by the medium. A high permittivity tends to reduce any electric field present.

pH: The degree of acidity or alkalinity of a solution. It is the negative logarithm of the hydrogen ion concentration in a solution, i.e. pH = $-\log_{10} [H^+]$. If the hydrogen ion concentration of a solution increases, the pH decreases, and vice versa.

Pharmacophore: A molecular description (that can be a three-dimensional structure) that carries (Greek: *phoros*) the essential steric and electronics features responsible for the biological activity of a drug (Greek: *pharmacon*). It does not represent a real molecule or a real association of functional groups, but an abstract model that accounts for the common molecular interaction capacities of a group of compounds towards their target structure.

Phosorylation: Biochemical reaction in which an organic substrate, for example a sugar or a protein, is combined with a phosphate ion by an enzymatic reaction.

Phospholipids: Main lipid component of cell membranes (q.v.). Phospholipids are heterogeneous molecules composed of glycerol, phosphate, two hydrophobic fatty acid tails, and »headgroups« with different chemical properties. Phospholipids form bilayers in cell membranes: the hydrophobic fatty acid residues pack against each other in a double layer, with the headgroups facing the exterior of the membrane.

Polymerase: An enzyme capable of synthesizing new strands of DNA from a single stranded template and free deoxynucleotides under appropriate reaction conditions.

Polymerization: A chemical reaction resulting in the bonding of small molecular units (monomers) to form a larger molecule (polymer).

Polysaccharide: A carbohydrate consisting of a large number of linked simple sugar, or monosaccharide, units. Examples of polysaccharides are cellulose and starch.

Primary structure: The linear amino acid sequence of the polypeptide chain including post-translational modifications and disulfide bonds.

Proteases (also called peptidases): Enzymes which break peptide bonds of proteins. The process is called proteolytic cleavage. They use a molecule of water for this and are thus classified as hydrolases.

Quaternary structure: The association of multiple polypeptide subunits to form a functional protein

Reading frame: The possible ways of reading a nucleotide sequence as a series of triplets translated into codons for protein synthesis.

Retrovirus: a virus (q.v.) which has a genome consisting of double-stranded RNA. It uses the reverse transcriptase enzyme to "retrotranscribe" its genome from RNA into DNA. The DNA is then inserted into the host's genome by an enzyme called integrase. The viral genome integrated into the host genome is called a provirus. Human immunodeficiency virus (HIV) is a retrovirus.

Ribosome: cellular particles made of protein and RNA which catalyze the synthesis of proteins according to the instructions contained in messenger RNA (mRNA) (q.v.). This process is called translation.

Rossman fold: Structural motif (q.v.) formed by three layers, two external α helical layers packing against a central parallel β sheet of six strands. It is the nucleotide binding motif found in nicotinamide adenine dinucleotide (NAD) binding proteins.

Secondary structure: Contiguous segment of amino acid residues with repeating values of the two angles ϕ and ψ. Very commonly observed secondary structures are α-helices and β-strands.

Server: A computer or software providing services to remote client machines or applications, such as supplying page contents (texts or other resources) or returning query results.

Speciation: The evolutionary development of a new species, usually as one population separates into two different populations no longer capable of producing fertile offspring when mating with each other.

Statistical distribution: A function describing the probability that a given value will occur.

String: A sequence of simple objects. They can be sequences of characters, tokens in a formal grammar, states in automata, DNA nucleotide sequences, protein amino acid sequences, bits, etc..

Structural genomics: The effort to determine the 3D structures of large numbers of proteins using both experimental techniques and modeling methods.

Structural motif: a three-dimensional arrangement of secondary structure elements of a protein chain, which appears in a variety of molecules.

Substrate: The molecule which is acted upon by an enzyme. The substrate binds at the enzyme's active site (q.v.), and is converted into products by a chemical reaction catalyzed by the enzyme.

Supersecondary structure: Combinations of two or three consecutive secondary structure elements in specific geometric arrangements observed in many different proteins.

Taxonomy: Classification according to a pre-determined system. In biology taxonomy classifies organisms according to their kingdom, phylum, class, order, family, genus, and species.

Taylor series: Series expansion of a function. The Taylor theorem states that any function satisfying certain conditions can be expressed as:

$$f(x) = \sum_{n=0}^{\infty} \frac{f^{(n)}(a)}{n!}(x-a)^n$$

For example, the Taylor series of the function sin(x) is:

$$\sin(x) = \sin(a) + \cos(a)\frac{(x-a)}{1!} - \sin(a)\frac{(x-a)^2}{2!} - \cos(a)\frac{(x-a)^3}{3!} + \cdots$$

because the derivative of sin(x) is cos(x) and the derivative of cos(x) is –sin(x). If we compute it for a = 0, we find:

$$\sin(x) = 0 + \frac{1}{1!}(x-0) - \frac{0}{2!}(x-0)^2 - \frac{1}{3!}(x-0)^3 + \cdots \approx x - \frac{x^3}{6} + \cdots$$

This means that the value of sin(x) can be approximated by

$$x - \frac{x^3}{6}.$$

Taking advantage of this approximation, the value of the sine of an angle, for example 18° = 0.314 rad, can be easily computed:

$$\sin(0.314) \approx 0.314 - \frac{0.314^3}{6} = 0.3088$$

(compared with the tabulated value of 0.3090).

Tertiary structure: The level of protein structure that describes the way in which the protein chain is folded into a specific three-dimensional arrangement.

TIM barrel: A protein topological arrangement consisting of eight parallel β strands connected to each other by α helices in such a way that the strands form a central barrel decorated with helices on their external surface.

Transfer RNA (abbreviated tRNA): tRNA is a small RNA chain (74–93 nucleotides) that transfers a specific amino acid to a growing polypeptide chain during translation. It has sites for amino-acid attachment and a specific codon (called anticodon) different for each tRNA. The anticodon contains the complementary bases to those coding for the amino acid to be transferred. It is the "adaptor" molecule hypothesized by Francis Crick, which mediates recognition of the codon sequence in mRNA (q.v.) and enables its translation into the appropriate amino acid.

Virus: A virus is a non-cellular entity composed merely of genetic material (DNA or RNA) surrounded by a protein envelope. Viruses can reproduce only within living cells into which they inject their genetic material. Some viruses can carry some essential proteins within their envelope.

World Wide Web: Created in 1989, the Web relies on an Internet standard (the hypertext transport protocol abbreviated http) which specifies how an application can locate and use resources stored on another computer on the Internet.

X-ray crystallography: X-ray crystallography is a technique in which the pattern produced by diffraction of X-rays through the lattice of atoms in a crystal is analyzed to deduce the structure of the lattice. X-ray photons interact with the electrons that surround the atoms, not with the atomic nuclei.

X-ray: Electromagnetic radiation of non-nuclear origin within the wavelength interval of 0.1 to 100 Å (between gamma-ray and ultra-violet radiation).

Index

a

α helix, see alpha helix
ab-initio prediction methods 70
acquired immunodeficiency syndrome, see AIDS
activation function, see transfer function
active site 3, 12–13, 27, 47–48, 112, 169, 172
AIDS 171, 174
alignment 37, 41, 46–47, 54, 75–78, 82, 86–87, 97–99, 108, 123, 125, 182
– global 79–80
– local 79, 94
– multiple 89, 91–96, 98, 110–112, 117, 121, 124, 131, 138, 151, 153, 156, 177–180
– score 80–81, 83–85, 92, 122
alpha helix 5–6, 8, 11–13, 16, 22, 43–44, 46, 64–65, 100, 131, 137, 140, 151, 160, 164, 166, 169
Anfinsen 24, 26
antibody, see immunoglobulin
antigen 3, 103
antigen binding loops, see antigen binding site
antigen binding site 102–104
architecture 11, 22–23, 35, 81, 118
ASTRAL 21
atomic solvation 121–122
automatic learning methods 137, 142, 153

b

β barrel, see beta barrel
β bulge, see beta bulge
β sheet, see beta sheet
β strand, see beta strand
back-propagation 144
background distribution 64, 83–86, 120
Bayes theorem 131
Bayesian logics 131, 134
benchmarking 50–51
beta barrel 10, 161, 166–167
beta bulge 44
beta sheet 6–8, 11–13, 43–44, 101, 169
beta strand 6–8, 22, 44, 46, 99, 101, 119, 131, 137, 140–141, 151, 160, 166
BLAST 75, 83–84, 86–87, 89–90, 96–97
– masking 85
BLOSUM 81–83, 154
Boltzmann distribution 63, 68
Boltzmann's law 63
Born-Oppenheimer approximation 56

c

canonical structure 103–104
CASP 52–54, 62, 70–71, 107, 109–110, 112–113, 121, 125, 129, 186–187
catalytic site, see active site
CATH 23
Chou and Fasman algorithm, see Chou and Fasman method
Chou and Fasman method 140, 142
circular dichroism spectroscopy 14
CLUSTAL 75, 93–94, 97
CNG 181–183
comparative modelling 53, 73 ff, 80, 84, 98–99, 108–109, 112–113, 117, 125–126, 170
core 17, 19, 36, 38–39, 47, 75–76, 81, 97–99, 107, 113, 119
correlated mutations 154, 156
Coulomb's law 58–60
covalent interactions 56
cumulative matrix 78, 80
cyclic nucleotide-gated channels, see CNG
cytokine 175

d

Darwin 29, 31
database search 75, 83–84, 90, 97, 151
– E-value 86–89, 91, 124

– p-value   86–87, 91
– score   86–87, 91, 96
Dayhoff, Margaret   81
dead end elimination   107
deletions   31, 35–37, 76–81, 94, 96–98, 108, 120, 122, 151
dielectric constant   58, 60–61
dihedral angle   1, 4–5, 7, 49, 57–58, 99, 106, 108, 135
dipole   60
dipole moment   61
distance geometry   138
distance map   22–23
distance matrix, see distance map
disulfide bond   12, 24–26
disulfide bridge   182
      – see also disulfide bond
domain   10–11, 23, 27, 36, 42, 53, 90–91, 98, 102, 112, 169, 173, 182
domain boundary   90
double bond   1–2, 56, 159
DSSP   44, 151

e
eigenpair   139
eigenvalue   138–139
eigenvector   138–139
electron density   15–17, 20–21, 56, 102
electrostatic interactions   58–59
energy landscape   35, 65
energy minimization   65–66, 110
enthalpy   24, 27
entropy   12, 24, 26–27
error space   146
error surface   145
EVA   51, 153
exon   36
extreme value distribution   86

f
false negatives, see FN
false positives, see FP
FASTA   75, 83–86, 89, 97
fibrinopeptides   32
fitness function   68–69, 107–108, 119
FN   87, 89
fold   10, 22, 27, 36, 73, 123–124, 126–127
fold recognition   53, 74, 108, 127, 131, 167, 175, 177
      – analogous   73, 117ff, 121, 125
      – homologous   73, 121, 125
Folding@home   70

FP   88–89, 91, 96
Fragfold   130–131, 135
fragment-based methods   74, 127ff
free energy   24, 26, 34–35, 55, 62
free energy of transfer   162–163
frozen approximation   122–123
FSSP   23

g
G-protein coupled receptors, see GPCR
gap penalty   83, 91, 94, 98, 120
Garnier-Osguthorpe Robson method, see GOR method
Gaussian distribution   85
gay related immunodeficiency disease, see AIDS
GB viruses   176–177
GBV-A, see GB viruses
GBV-B, see GB viruses
GBV-C, see GB viruses
GDT-TS   48–49, 54, 109, 186
genetic algorithms   67–69
genetic code   29–30
glycerol   160
glycoprotein   172, 176, 178
GOR method   142
GPCR   185

h
HCV   176–181, 183–184
helical wheel   165
hemoglobin   32–34, 117
hepatitis C virus, see HCV
HGV   177
hidden Markov models   89, 94–97, 156, 166–167
High-throughput screening   183
HIV   171–172, 174
HMM, see hidden Markov models
homologous proteins   27, 35, 39, 50, 84, 87, 97, 107–108, 117, 124, 147, 151
homology   23, 34
homology modelling, see comparative modelling
Hooke's law   56
hydrogen bond   5ff, 12–13, 17, 24, 43–44, 59, 66, 99–101, 103–104, 134, 160–161, 163
hydrophobic moment   165–166
hydrophobicity   28
      – hydrophobicity plot   164–166
      – hydrophobicity scale   163

i

immunoglobulin 3–4, 99, 101–106, 117
insertions 31–32, 35–37, 76–81, 94, 96–98, 108, 120, 122, 151
intron 36

j

Jack knive 50

k

kernel function 148–149
knowledge based potentials, see statistical mechanics potentials

l

lactalbumin 171
lead compound 183
leptin 174–176
leptin receptor 175–176
Levinthal paradox 26–27
Livebench 51
long-range contacts 153, 156
loop 8, 12–13, 44, 46, 48, 98–107, 110, 141, 167
lysozyme 169–170

m

meiosis 31–32
membrane 23, 117, 159–163
membrane proteins 159–161, 164, 166–167, 172, 185
Mendel 29
mitosis 31
model accuracy 41, 47, 76–77, 109, 126, 130, 180–181, 186
model optimization 107, 124
Modeller 108
module 11
molecular dynamics 66–67, 70, 110, 167
molecular replacement (in X-ray crystallography) 15, 186–187
Monte Carlo 67–68, 122, 156
moonlight proteins 28
motif 10, 27, 128
mutation 31–32, 34–36, 68, 76, 79, 81, 151, 154, 176, 180
    – missense 31
    – nonsense 31
myoglobin 33–34, 169–170

n

natively unfolded proteins 28
Needleman and Wunsch algorithm 78–80
neural network 142–144, 147–148, 150–151, 153, 162, 167
neuron 143–144, 147, 152
new fold prediction 73
Newton equation 66
Newton's law 55
NMR 14, 17–20, 22, 52, 138, 184
node (in neural networks), see neuron
nuclear magnetic resonance, see NMR
nucleation site 141

o

orthology 34, 84

p

pair interaction potential, see statistical mechanics potential
pair potential, see statistical mechanics potential
PAM 81–83, 154–155
Paracelsus challenge 36
paralogy 34, 84
partition coefficient 162
Pauli exclusion principle 61
PDB 10–11, 20–22, 25, 35, 42, 51, 74–75, 97
pepsin 2–3, 172
peptide bond 1–4, 13, 63, 104
permittivity 58
PFAM 97
pharmacophore 183
PHD 151–153
ϕ angle 6–8, 12–13, 43–44, 69
phi angle, see ϕ angle
phospholipids 159–160
phylogeny 32
potential energy 62, 65–67, 135
prediction server 51, 110, 135
primary structure 4, 100
prior probability 131, 133–134
profile 89, 94, 96, 121, 124, 132, 151, 153
profile-based methods 119–120
profile–profile methods 124
protease 3, 12–14, 27–28, 170–174, 176, 178–180
Protein Data Bank, see PDB
protein design 135
protein kinase 13, 184
pseudo-counts 94
ψ angle 6–8, 12–13, 43–44, 69
psi angle, see ψ angle
PSI-BLAST 75, 83, 96–97, 124, 151, 153
PSIPRED 153

q

Q index (in secondary structure prediction) 44
Q3 index (in secondary structure prediction) 44
quaternary structure 4, 27, 178

r

radius of gyration 134
Ramachandran plot 7–8
random coil 24, 28
random distribution, see background distribution
Receiving Operator Curve, see ROC
replacement (of amino acids), see mutation
retrovirus 171
rhodopsin 164, 185
rmsd 19, 37–39, 47–48, 170
RNase A 25–26
ROC 89
root mean square deviation, see rmsd
Rosetta 130, 132, 135–136
Rossman fold 9–10, 117
rotamer libraries 106–107

s

sampling 145
Schrödinger equation 55–56, 58
SCOP 23
SCWRL 107
SDR, see structurally divergent regions
secondary structure 4, 7, 11, 17, 20, 22–23, 36, 38, 42, 46, 51, 59, 64, 81, 98–99, 102, 117–119, 135, 137, 141, 160, 166–167
– prediction 41, 43–44, 120, 124, 138, 140, 144, 147, 150–153, 161, 176–177
Segment Overlap measure, see SOV
Semliki Forest virus, see SFV
sensitivity 87–89
sequence alignment, see alignment
SFV 178–179
side chain 6–7, 12–14, 17, 25, 27, 49, 75–77, 99–100, 103, 106–108, 110, 113, 121, 154, 186
similarity matrices, see substitution matrices
simulated annealing 68, 108, 135
single bond 1, 56
sliding window 151–152, 163–164
Smith and Waterman algorithm 80
solvation energy 121
solvation free energy 121
solvation potential 121
solvent accessibility 51, 119–121, 134

SOV 44–46
sparse coding 150
specificity 87–89
spin 18
statistical mechanics potentials 62–65, 107–108, 121–122, 131, 135
stems 101–102
STRIDE 44, 151
structural genomics 27, 29
structural superposition 23, 37, 46–47, 51, 96, 98
structurally divergent regions 75, 98–99, 135
substitution, see mutation
substitution matrices 81–83, 86, 98
sum of pairs score 93
superfolds 117, 119
superposition, see structural superposition
supersecondary structure 9–10, 127, 131
supervised learning, see supervised training
supervised training 144
support vector machines 148–150, 162
SVM, see support vector machines
SwissModel 98

t

T-COFFEE 75, 93–94, 97
target function, see fitness function
Taylor power series 57
TBEV 178–179
template 77, 84, 91, 96–99, 101, 108–112, 121–124, 126–127, 131, 170, 187
tertiary structure 4, 8, 26, 46–47, 49
test set 50, 147–148, 151
threading 119, 121–123, 175
Tick Borne encephalitis virus, see TBEV
TIM barrel 11, 117
time step (in molecular dynamics) 66–67
TMHMM 166
TN 88–89
topology 11, 23, 27, 36, 118, 167
TP 87, 89
training set 50–51, 142, 146–147, 151
transfer function 143–144, 147
transmembrane proteins, see membrane proteins
tree 32, 34, 94
triangle inequalities 139
true negatives, see TN
true positives, see TP
trypsin 2–3, 21
turn, see loop

u

unfolded conformation, see random coil
unfolded protein, see random coil
unsupervised learning, see unsupervised training
unsupervised training 144

v

validation set 146–148, 151
van der Waals potential 24
van der Waals radius 61
virtual screening 183–186

w

water 24–25, 59, 121

x

X-ray crystallography 14–17, 19, 52, 61, 101, 110, 169–170, 181, 184, 186
– B factor 17–18, 21–22, 49
– Debye-Waller factor, see B factor
– occupancy 17, 21
– R factor 17, 20, 22
– resolution 17, 22
– Rfree factor 17, 20
– temperature factor, see B factor

z

Z-score 123